Praise for the fu

"Indispensa

"An exceptional reference work!" —*America*

"Excellent, scholarly yet readable treatment of the Narnia stories." —*The Reformed Journal*

"The clear prose, accurate scholarship, plentiful cross-references and charming new illustrations . . . may help older readers recapture a shadow of their first encounter with the texts themselves." —*Los Angeles Times*

"Ford views C. S. Lewis's *Chronicles of Narnia* as a source of both illustration and assistance on Christian teaching for readers of all ages; and his footnotes are rich in scriptural and theological parallels." —*Library Journal*

"The book, with its charming illustrations by Lorinda Bryan Cauley, is plain fun, whether it is used as a reference work or studied from cover to cover." —Chad Walsh

"This is an excellent book to add to church and school libraries, although it could also be a worthwhile reference work for families with 'middle-aged children.'" —*The Banner*

"Supplementing, supplanting, and/or fulfilling all previous works on the Narnian *Chronicles*, this superb book is sure to stand as a classic in its field." —*Mythlore*

Pocket Companion to Narnia

Eustace, Edmund, and Lucy bid farewell to Reepicheep, the courageous mouse, as his coracle skims the smooth green current and he rides the wave of eternity into Aslan's country. (VDT 16)

Pocket Companion to Narnia

A Guide to the Magical World of C. S. Lewis

Paul F. Ford

Illustrated by Lorinda Bryan Cauley

HarperSanFrancisco
A Division of HarperCollins*Publishers*

Book design by rlf design

Diagrams by Dr. Stephen Yandell have been redrawn by Lydia Hess.

Library of Congress Cataloging-in Publication Data is available.
ISBN-10 0-06-079128-4

05 06 07 08 09 WEBCOM 10 9 8 7 6 5 4 3 2 1

for

Paul Ford

(my namesake)

his brother David and his sister Emily

his mother Laura and his father Raymond (my brother)

John Lake

(who dedicated his first "book" to me)

his brothers Erik and Thomas

his mother Silvia and his father Don

Nathan Blackmon

his brothers Andrew and Alex

his mother Sheri and his father Rick

Becky Cerling and Joshua Falconer,

without whom . . .

all, my companions in Narnia and in life

Contents

Abbreviations

The books of the *Chronicles of Narnia* are abbreviated as follows:

LWW	*The Lion, the Witch and the Wardrobe*
PC	*Prince Caspian*
VDT	*The Voyage of the* Dawn Treader
HHB	*The Horse and His Boy*
SC	*The Silver Chair*
MN	*The Magician's Nephew*
LB	*The Last Battle*
[SPOILER(S)]	Don't read the next sentence (paragraph) because it tells you something very important in a story you may not have read yet.
N.Y.	Following a date means "Narnian Years"; thus 1014 N.Y. is Narnian time. See *The Land of Narnia*, 31.

List of Maps, Illustrations, and Diagrams

Maps

Illustrations

Diagrams

Using the Pocket Companion

A simplified version of this section is available for printing as a bookmark at my website: www.pford.stjohnsem.edu/ford/cslewis/narnia.htm.

Where to Begin

You have begun in the right place by coming to this page. If you have not read the *Chronicles,* read them first (in the order "245-3617"—see below, 14–15).

After you have read each book, then come back to this page and read these few pages called "Using the *Pocket Companion.*" Then read "Advice to an Intelligent Reader from an Intelligent Reader *or* Reading This Will Make You Smarter."

For a more directed study of the *Chronicles,* see "Guide to the Most Important Entries" at the end of this section. Reading any of the "Major Characters and Items" (listed there) will also lead you into the heart of this *Pocket Companion.*

How to Find an Entry

Entries are in alphabetical order. Characters are listed by first name (e.g., Andrew Ketterley is listed under "A" for Andrew), and honorifics or titles are left off (e.g., Prince Caspian is listed under "C" for Caspian, and Mr. Beaver is listed under "B" for Beaver, Mr.).

How to Read a Sample Entry

ENTRY TITLES — Abbreviated book titles and chapter numbers indicate where the topic under discussion can be found in the *Chronicles*. For instance, "SC 3 and 4" refers you to chapters 3 and 4 of *The Silver Chair.* SMALL CAPITALS are used to alert you to a cross reference, namely, a separate entry on this person, place, thing, or idea.[1] Singular and plural verb and noun forms of the entry title will both be set in small capitals (e.g., ANIMAL and ANIMALS, PRAY and PRAYER). Some longer titles, such as SEVEN FRIENDS OF NARNIA and CASTLE OF THE WHITE WITCH, are listed under the first word of that phrase, not the last (i.e., "Seven ..." and "Castle ..."). On those rare occasions when two cross references are run together, an asterisk functions as a divider between these cross references. For example, under ADULTS, the phrase "Digory recognizes right away that Uncle ANDREW is an EVIL* MAGICIAN" indicates that you

[1]Notes are listed following the entry, except here and in *Advice to an Intelligent Reader.*

will find more information at ANDREW, at EVIL, and at MAGICIAN. "[SPOILER(S)]" means "Don't read the next sentence/paragraph if you have not yet read the book in question because it tells you something very important to the story."

At the end of some entries you will find cross references not specifically mentioned in the text listed in [brackets].

[ASTRONOMY, NARNIAN; PROVIDENCE.]

Bibliography

I use the following short titles and abbreviations in the notes. Here is more information:

The Abolition of Man
(first published in 1946; now available from HarperSanFrancisco, 2001)

Biography
Roger Lancelyn Green and Walter Hooper, *C. S. Lewis: A Biography* (New York: Harcourt Brace Jovanovich, 1974, and London: Collins, 1974)

Christian Reflections
Walter Hooper, ed. (Grand Rapids: Eerdmans, 1967)

The Discarded Image
(London: Cambridge University Press, 1964)

Essays Presented to Charles Williams
(Oxford: Oxford University Press, 1947; Grand Rapids: Eerdmans, 1966)

The Four Loves
(New York: Harcourt Brace Jovanovich, 1960)

God in the Dock
Walter Hooper, ed. (Grand Rapids: Eerdmans, 1970)

The Great Divorce
(first published in 1946; now available from HarperSanFrancisco, 2001)

Hooper
Walter Hooper, *C. S. Lewis: A Companion and Guide* (London: HarperCollins, 1996)

The Land of Narnia
Brian Sibley, *The Land of Narnia: Brian Sibley Explores the World of C. S. Lewis* (New York: HarperTrophy, 1998)

Letters to an American Lady
Clyde S. Kilby, ed. (Grand Rapids: Eerdmans, 1967)

Letters to Children
C. S. Lewis Letters to Children, Lyle K. Dorsett and Marjorie Lamp Mead, eds. (New York: Macmillan, 1985; Scribner, 1996)

Letters to Malcolm: Chiefly on Prayer
(New York: Harcourt, Brace & World, 1964)

Letters II
C. S. Lewis: Collected Letters, Volume II: Books, Broadcasts and War 1931–1949 (London: HarperCollins, 2004)

Letters 1988

Letters of C. S. Lewis, W. H. Lewis, ed., revised and enlarged by Walter Hooper (San Diego: Harcourt Brace, 1988)

Mere Christianity

(first published in 1952; now available from HarperSanFrancisco, 2001)

Miracles

(first published in 1947; revised 1960; now available from HarperSanFrancisco, 2001)

Of Other Worlds

Walter Hooper, ed. (New York: Harcourt Brace & World, 1966)

Past Watchful Dragons

Walter Hooper (New York: Macmillan, 1979)

Perelandra

(first published in 1943; latest edition from Scribner, 2003)

The Problem of Pain

(first published in 1940; now available from HarperSanFrancisco, 2001)

The Screwtape Letters

(first published in 1942; now available from HarperSanFrancisco, 2001)

Surprised by Joy

(New York: Harcourt, Brace & World, 1955)

The Weight of Glory
(first published in 1949; revised in 1980; now available from HarperSanFrancisco, 2001)

The World's Last Night
(New York: Harcourt, Brace & World, 1960)

Variants

The most significant differences between the pre-1994 American and the British editions (now the English editions for the world) of the *Chronicles* are noted in the entries DREAM(S), FENRIS ULF, SECRET HILL, and WORLD ASH TREE.

In this *Pocket Companion to Narnia*, "American edition(s)" = Any edition of the *Chronicles of Narnia* typeset and printed in the United States of America before 1994. These editions have the changes Lewis made after he had proofread the second set of galleys sent to him by his American publisher. Since he rarely made any changes, these are significant. (See *Advice to an Intelligent Reader*, 11.)

"British edition(s)" = Any edition of the *Chronicles of Narnia* typeset and printed in Great Britain before 1994 and any edition in the English language published anywhere in the world after 1994; after this date the volumes are renumbered according to the order of the internal chronology. These editions do not contain the changes Lewis made after he had proofread the second set of galleys sent to him by his American publisher.

Guide to the Most Important Entries

Major Characters and Places. The following entries are major characters and places and are a good place to start your reading of this *Pocket Companion to Narnia.*

> ARAVIS, ASLAN, BREE, CAIR PARAVEL, CASPIAN X (PRINCE CASPIAN), DIGORY KIRKE, EDMUND PEVENSIE, EUSTACE CLARENCE SCRUBB, HWIN, JILL POLE, LUCY PEVENSIE, NARNIA, PETER PEVENSIE, POLLY PLUMMER, PUDDLEGLUM, REEPICHEEP, SHASTA, SUSAN PEVENSIE, WHITE WITCH.

Another way to begin is by consulting the following essential entries.

For the *Chronicles of Narnia* in general:

> *Using the* Pocket Companion, *Advice to an Intelligent Reader,* ASLAN; AUTOBIOGRAPHICAL ALLUSION(S); BIBLICAL ALLUSION(S); FEELING(S); IMAGINATION; LITERARY ALLUSION(S); MAGIC; MYTHOLOGY; PROFESSION(S) OF FAITH; STOCK RESPONSE(S).
>
> If you have time, see also: ADULT(S); ANIMAL(S); AUTOBIOGRAPHICAL ALLUSION(S); CHILDREN; "FOR THE FIRST TIME"; GEOGRAPHY, NARNIAN (plus the maps in the middle of the book); KING(S), QUEEN(S); PAIN; PLATO; PRAYER; PROVIDENCE; SILENCE; SLEEP; SMELL(S); SOUND(S); YOUTH; and *Appendices One* and *Two.*

For Book One: *The Lion, the Witch and the Wardrobe*

ASLAN'S NAME; ASLAN'S VOICE; CENTAURS; DEPRAVITY; DOMESTICITY; DRYAD(S); EDMUND PEVENSIE; EMPEROR-BEYOND-THE-SEA; HIERARCHY; KING(S), QUEEN(S); LUCY PEVENSIE; MAGIC; NUMINOUS; PETER PEVENSIE; PLATO; SON OF ADAM, DAUGHTER OF EVE; SUSAN PEVENSIE; TREE(S); TUMNUS; WHITE WITCH.

For Book Two: *Prince Caspian: The Return to Narnia*

ASTRONOMY, NARNIAN; CASPIAN X; COURTESY; DANCE; DWARF(S); FEAR; GREATNESS OF GOD; HONOR; HOPE(S); LONGING; PRACTICAL NOTES; REVELRY; ROBE(S); SCHOOL(S); TELMAR; TIME; TRANSITION(S); TRUFFLEHUNTER; TRUMPKIN.

For Book Three: *The Voyage of the* Dawn Treader

ADVENTURE; ASLAN'S COUNTRY; CORIAKIN; COURAGE; COWARDICE; DARK ISLAND; *DAWN TREADER*, DAWN TREADER, DREAM(S); DUFFER(S); EUSTACE CLARENCE SCRUBB; PRIVACY; REEPICHEEP; SILENCE; STAR(S); VANITY.

For Book Four: *The Horse and His Boy*

ARAVIS; BREE; COMFORT; CRY(ING); HORSE(S); HWIN; OBEDIENCE; RABADASH; SEXISM; SHASTA; TRINITY; YOUTH.

For Book Five: *The Silver Chair*

CRY(ING); JILL POLE; MEMORY; PUDDLEGLUM; QUEEN OF UNDERLAND; QUEST; REDUCTIONISM; RILIAN; UNDERLAND; YOUTH.

For Book Six: *The Magician's Nephew*

ANDREW KETTERLEY; CHARN; CURIOSITY; DIGORY KIRKE; GREED; HUMOR; POLLY PLUMMER; POSITIVITY; RIGHT AND WRONG; WOOD BETWEEN THE WORLDS.

For Book Seven: *The Last Battle*

DEATH; DOG(S); DOOR(S); EMETH; ESCHATOLOGY; HISTORY; JEWEL; JOY; JUDGMENT; LAST BATTLE(S); PUZZLE; RAILWAY ACCIDENT; SHIFT; STABLE; TASH; TECHNOLOGY; TERM-TIME; TIRIAN; UNIVERSALISM.

Advice to an Intelligent Reader from an Intelligent Reader or *Reading This Will Make You Smarter*

This book has many spoilers! Please read the *Chronicles of Narnia* first. Only then read this *Pocket Companion.* I can't be your companion until you have "gone there" first.

C. S. Lewis wrote these books for you, and he meant you to read them with your head and your heart and your feelings. The books are so exciting that you will probably race through them the first time. OK.

But when you reread them, pay attention to the color words, the verbs, the adjectives, and the adverbs.

Try to avoid looking for hidden meanings, as if the *Chronicles* were codebooks.

C. S. Lewis was very careful not to decode the *Chronicles* for the children who wrote him about their meaning. Typical of his responses is the answer he made to Hila Newman when she wrote to him about the meaning of the ending of *The Voyage of the* Dawn Treader:

> As to Aslan's other name, well I want you to guess. Has there never been anyone in this world who (1) Arrived at the same time as Father Christmas (2) Said he was the son of the Great Emperor (3) Gave himself up for someone else's fault to be jeered at and killed by

wicked people (4) Came to life again (5) Is sometimes spoken of as a Lamb (see the end of The Dawn Treader). Don't you really know His name in this world? Think it over and let me know your answer! (*Letters to Children*, 3 June 1953)

In Which Order Should the *Chronicles* Be Read?

The Order by Year of First Publication—the Canonical Order

Before 1994 there was only one reading order, what scholars have come to call *the canonical order*, the order in which the chronicles were first published, and they were numbered accordingly:

The Lion, the Witch and the Wardrobe (published 1950)
Prince Caspian: The Return to Narnia (published 1951) — the Caspian triad
The Voyage of the Dawn Treader (published 1952) — the Caspian triad
The Silver Chair (published 1953) — the Caspian triad
The Horse and His Boy (published 1954)
The Magician's Nephew (published 1955)
The Last Battle (published 1956)

So, although C. S. Lewis completed *The Horse and His Boy* before *The Silver Chair,* the latter was published first, in order to keep together the triad of books in which Caspian is a major character.

The Order by Internal Chronology—the Chronological Order

Very quickly, readers noticed that the stories were not in the order of their internal chronology, and some wrote Lewis about the issue. Here is a typical reply, written on 23 April 1957 but not published until 1985:

> Dear Laurence
>
> I think I agree with your order for reading the books more than with your mother's.[1] The series was not planned beforehand as she thinks. When I wrote *The Lion [, the Witch and the Wardrobe]* I did not know I was going to write any more. Then I wrote *P.[rince] Caspian* as a sequel and still didn't think there would be any more, and when I had done *The Voyage [of the* Dawn Treader] I felt quite sure it would be the last. But I found I was wrong. So perhaps it does not matter very much in which order anyone reads them. I'm not even sure that all the others were written in the same order in which they were published. I never keep notes of that sort of thing and never remember dates.

[1]Laurence's mother felt that the seven chronicles of Narnia should be read in the order in which they were published, since she assumed that this sequence was intentional. Laurence, however, believed that the stories should be read chronologically according to Narnian time: *The Magician's Nephew, The Lion, the Witch and the Wardrobe, The Horse and His Boy, Prince Caspian, The Voyage of the* Dawn Treader, *The Silver Chair,* and *The Last Battle*.

This letter suggested the change that HarperCollins implemented in 1994 when the decision was made to make the British editions the world English editions. This has come to be called the chronological order.

The Magician's Nephew
The Lion, the Witch and the Wardrobe
The Horse and His Boy
Prince Caspian
The Voyage of the Dawn Treader
The Silver Chair
The Last Battle

Most scholars and fans disagree with this decision and find it the least faithful to Lewis's deepest intentions and to the meanings the books have acquired among the most devoted readers.

The Best Reading Order—the Order by Essential Completion by Lewis Himself "Remember 245-3617"

The *Pocket Companion to Narnia* studies the *Chronicles* in a slight adaptation of the canonical order, what I have called the "order by essential completion." This order makes possible the reading of the book with the heart by attending to them in the order in which they came to Lewis's own heart and mind.

The Lion, the Witch and the Wardrobe (spring 1949)
Prince Caspian: The Return to Narnia (fall 1949)
The Voyage of the Dawn Treader (winter 1950)
The Horse and His Boy (spring 1950)

The Silver Chair (spring 1951)
The Magician's Nephew (fall 1951)
The Last Battle (spring 1953)

I urge all readers of the *Chronicles of Narnia* to "Remember '245-3617,'" that is, rearrange the post-1994 HarperCollins reordering before you read the books!

My worry is that the decision to reorder the books chronologically diminishes their impact on future readers, indeed, impedes readers from moving from *The Magician's Nephew* to *The Lion, the Witch and the Wardrobe* and thus to the end, which is "only the beginning of the real story."

Consider how *The Lion, the Witch and the Wardrobe* introduces the mystery of a world within a wardrobe and builds to the first use of the name "Narnia" in paragraph 10 of Chapter 2 and the revelation of Aslan in Chapter 7, "A Day with the Beavers." Contrariwise, *The Magician's Nephew* plops the reader unmysteriously into the plot of the whole series, using "Narnia" as the fortieth word a reader will now encounter.

But the pivotal insight that clinches the argument is found in the scene just cited: "None of the children knew who Aslan was *any more than you do;* but the moment the Beaver had spoken these words everyone felt quite different."[2] The five words I have emphasized show that we must read the books in the order in which they first came to the attention of the world of readers and rereaders, in the order in which the *meaning* of these glorious books grew beyond Lewis's late-formed *intention* to revise them. His

---

[2]LWW 8.

intention to emend the books—agreed to just two days before he died—is inferior to his attention to their meaning and their success at that level (his deeper intention).

If Lewis had been able to complete his intended revision, perhaps the chronological order would be the better. But the "245-3617 order" or the canonical order is to be preferred. Why? Helping to avoid the ever-present danger of decoding the *Chronicles,* these orderings carry the reader along in a less logical, less factual mode and present the pictures and the meanings of Lewis's stories in the way he first decided to tell them and in the way the first readers of the *Chronicles* enjoyed them.

How C. S. Lewis Wrote the *Chronicles of Narnia*

You can save the following, final piece of advice for reading later. It helps you get an overview of what C. S. Lewis was doing. Warning: SPOILERS ahead!

It is important to be aware of the fact that, unlike J. K. Rowling and her Harry Potter series, Lewis did not begin with any plan to write seven chronicles of Narnia.

Lewis seems to have worked out of four separate spurts of creative energy left over from his having just finished a very important book of philosophy and theology called *Miracles.* The first spurt of energy produced *The Lion, the Witch and the Wardrobe* and a rough draft of *The Magician's Nephew*, named "The Lefay Fragment."[3] The second re-

[3]See *Past Watchful Dragons*, 48–67.

sulted in *Prince Caspian: The Return to Narnia* and *The Voyage of the* Dawn Treader. The third was responsible for *The Horse and His Boy*. And the final impulse saw *The Silver Chair, The Last Battle*, and *The Magician's Nephew* (finished in *that* order).

Lewis's First Burst of Narnian Creativity (summer of 1948 to summer of 1949)

It could very well be that finishing *Miracles* and starting his autobiography, *Surprised by Joy,* in the spring of 1948, released in Lewis the energy to return to a book he had started and left off at the beginning of World War II with the arrival of the three girls evacuated from London.[4] *The Lion, the Witch and the Wardrobe* is caught up in describing the new world of Narnia, its enchantment by the White Witch, the lot of the Talking Beasts and the mythological creatures under her rule, the transformation of the country by the intervention of the Great Lion, and its subsequent rule by the four English children. The "Lefay Fragment"[5] seems to convey that Lewis wasn't yet ready to write the creation story of Narnia "because he had not lived long enough in the world and with the characters he had created"[6] and he wasn't far enough along in the understanding of the meaning of his own life story to appreciate what was

---

[4]See AUTOBIOGRAPHICAL ALLUSION(S).

[5]A two-chapter draft of what later became *The Magician's Nephew*. It is published in *Past Watchful Dragons*.

[6]Paul F. Ford, "The Chronicles of Narnia," in Jeffrey D. Schultz and John G. West Jr., eds., *The C. S. Lewis Reader's Encyclopedia* (Grand Rapids: Zondervan, 1998), 122.

happening to him in his own writing. He set aside this rough draft of *The Magician's Nephew* and paused.

Lewis's Second Burst of Narnian Creativity (summer of 1949 to fall of 1950)

The subtitle of *Prince Caspian*, "The Return to Narnia," provides the most important clue as to Lewis's intended meanings for the work (no other chronicle has a subtitle). The subtitle suggests that the book can be seen as a rough draft of *The Last Battle* in that both books ask the questions "Is the Christian story real now or only something that may have happened long ago?" and "What does the effect of the passage of time have on the reality and experience of faith?" (*Prince Caspian* twice refers to the events of *The Lion, the Witch and the Wardrobe* as "the golden age.") In *Prince Caspian* there are also pre-echoes of a major theme in both *The Last Battle* and also *The Magician's Nephew:* the right and wrong uses of nature and people.[7] *Prince Caspian* concentrates even more on the geography, astronomy, and history of Narnia and adds considerably to the cast of characters. Aslan sends the boys and Trumpkin to war alongside Caspian while he and the girls will liberate Narnia through the ecstatic.[8]

Then, in almost an excess of imaginative energy, Lewis

---

[7]The way *Prince Caspian* handles this theme is akin to J. R. R. Tolkien's chapter "The Scouring of the Shire" in Book Six of *The Lord of the Rings*. See Paul F. Ford, "*Prince Caspian*," in *The C. S. Lewis Reader's Encyclopedia*, 337.

[8]See Paul F. Ford, "*Prince Caspian*," in *The C. S. Lewis Reader's Encyclopedia*, 338.

wrote—in less than three months—*The Voyage of the* Dawn Treader, which takes the reader from the slave-trading intrigues on islands close to Narnia to the very threshold of Aslan's country in a high point of what we would have to call mysticism.[9] Lewis intended to end the *Chronicles* with *The Voyage of the* Dawn Treader, as you can see in the last scene of the book.

Lewis made the most significant change in all of the *Chronicles* when he revised the ending of Chapter XII of *The Voyage of the* Dawn Treader, "The Dark Island." Between the time he corrected the galley proofs of the British edition and the time he corrected the galley proofs of the American edition, he delivered the address "On Three Ways of Writing for Children" to the British Library Association. When someone objected that fairy tales are too violent for children, Lewis said: "I suffered too much from nightfears myself in childhood to undervalue this objection. I would not wish to heat the fires of that private hell for any child." He seems to have had second thoughts about whether "The Dark Island" was too frightening[10] and made the changes detailed in the entry DREAM(S).

Lewis's Third Burst of Narnian Creativity (spring and summer of 1950)

After a pause of at least two months, Lewis, like a portraitist finished with his main subject, started *The Horse and*

---

[9]See Paul F. Ford, "*The Voyage of the* Dawn Treader," in *The C. S. Lewis Reader's Encyclopedia,* 418–420.

[10]*Letters to Children,* 14 September 1953.

His Boy to fill in the background by telling a story that takes place in Calormen and Archenland.[11] Along the way, however, much as Shasta encounters Aslan in the fog (Chapter XI, a foreshadowing of Emeth in *The Last Battle*), Lewis was also overwhelmed.

Remember, Lewis was writing his autobiography at the same time. Imagine the effect on Lewis of taking the invitation Aslan makes to Shasta, "Tell me your sorrows," as an invitation addressed to himself. What a process for all of us readers, and what a climax in the vision of Aslan that rewards Shasta's honest dialogue!

Lewis's Final Burst of Narnian Creativity (summer of 1950 to spring of 1953)

So, after a second pause, of at least four months, Lewis plunged into the creation of the last three chronicles. He finished *The Silver Chair* first, labored at *The Magician's Nephew,* leaving it with some threads untied, and went on to start and complete *The Last Battle* before finally completing *The Magician's Nephew*.

The tone has changed when we enter the world of *The Silver Chair.*[12] The scene is Experiment House, the progressive school that Eustace Scrubb attends. In this fifth chronicle (published fourth), Lewis shows how aware he was of the two senses of the word *spell.* Tolkien had written,

---

[11]See Paul F. Ford, "*The Horse and His Boy*," in *The C. S. Lewis Reader's Encyclopedia,* 208–210.

[12]See Paul F. Ford, "*The Silver Chair*," in *The C. S. Lewis Reader's Encyclopedia,* 377–379.

"Small wonder that *spell* means both a story told, and a formula of power over living men."[13] Since the time he wrote *The Pilgrim's Regress,* Lewis had been alive to the paralyzing hold the enchantment of the spirit of the age has on the minds and hearts of living men and women. And it is this tone we detect very strongly in *The Silver Chair.* The emphasis is on discipline ("Remember, remember the Signs"), on the power of fear and of the desire for pleasure, and on obstinacy in belief.

In *The Magician's Nephew,*[14] Lewis returns to the threads of a story about the creation of Narnia that he had begun and almost immediately abandoned after he had finished *The Lion, the Witch and the Wardrobe.* Alive to his own memories of the turn of the century, he recreates those times as the background for a discussion of the consequences of the unbridled desire for knowledge and power in Andrew Ketterley and his nephew Digory. Jadis is menacingly real, as well as the logical outcome of what exists in its beginning stages in the boy and is well advanced in the uncle.

The Last Battle has all the quality of the "Twilight of the Gods" (one of Lewis's earliest experiences) transformed, though not right away, by the Christian hope of *eucatastrophe,* Tolkien's term for the "joyous sudden turn" of a fairy tale.[15] The story has a life of its own and it moves easily through the eschatological themes of death, judgment,

---

[13]*Essays Presented to Charles Williams,* 56.

[14]See Paul F. Ford, "*The Magician's Nephew,*" in *The C. S. Lewis Reader's Encyclopedia,* 262–264.

[15]*Essays Presented to Charles Williams,* 81.

hell, and heaven. It is not only the fitting conclusion to the *Chronicles:* Given Lewis's Christian faith, it is their only possible conclusion.[16]

In summary, the Narniad is a seven-volume magician's book devised by Lewis to break bad enchantments and bring about re-enchantments by reawakening a longing for Aslan and Aslan's country. Lewis spoke in "The Weight of Glory": "Do you think I am trying to weave a spell? Perhaps I am; but remember your fairy tales. Spells are used for breaking enchantments as well as for inducing them. And you and I have need of the strongest spell that can be found to wake us from the evil enchantment of worldliness which has been laid upon us for nearly a hundred years."[17]

Sources for Further Help

The advent of the Internet has made many Narnia-related resources available. The best Lewis site to date is http://cslewis.drzeus.net/. The entire contents of the *Chronicles* are searchable at www.amazon.com under the single-volume paperback edition of the *Chronicles,* ISBN 0–06–623850–1; click on "search inside this book."

Updates to this *Pocket Companion* and to the *Companion*

---

[16]See Paul F. Ford, *"The Last Battle,"* in *The C. S. Lewis Reader's Encyclopedia,* 232–233.

[17]*The Weight of Glory,* ¶5. For a discussion of magic as curiosity, or an immoderate striving for knowledge, see Josef Pieper, *The Four Cardinal Virtues* (Notre Dame: University of Notre Dame Press, 1966), 199.

and other Narnian helps are available at my website: www.pford.stjohnsem.edu/ford/cslewis/narnia.htm. Contact me at companion_to_narnia@hotmail.com.

I urge my traveling companions to Narnia to continue to point out to me features of this marvelous land I might more clearly see, characters in this wonderful country I might more deeply understand, and events in these great stories I might more insightfully interpret.

—Paul F. Ford
Camarillo, California
March 25, 2005

The Anniversary of the Day the
Ring of Power Was Destroyed in
the Furnace of Mount Doom

ADOLESCENCE — See YOUTH.

ADULT(S) — Throughout the *Chronicles,* Lewis uses "grown-up" as a synonym for narrow-minded, unimaginative, and too practical thinking (see PRACTICAL NOTES for the good side of practicality). LUCY identifies as "grown-up" the skepticism she sees in SUSAN's question "Where do you think you saw ASLAN?" (PC 9). Lewis comments that "it is the stupidest grown-ups who are the most grown-up" (SC 16). Also, SHASTA "has the fixed habit of never telling grown-ups anything if he [can] help it" (HHB 5). It is this kind of adult thinking that created EXPERIMENT HOUSE. Perhaps the main complaint CHILDREN and YOUTH have against grown-ups is that they have lost their IMAGINATIONS. One short conversation between POLLY and DIGORY illustrates this quite well. When they propose to explore the long-unoccupied house beyond Digory's, Digory says, "It's all rot to say a house would be empty all those years unless there was some mystery." "Daddy thought it must be the drains," replies Polly. Digory comments, "Pooh! Grown-ups are always thinking of uninteresting explanations" (MN 1).

Among the good human grown-ups of the *Chronicles* are King CASPIAN X, Professor Digory Kirke, DRINIAN, ERLIAN, King FRANK, Queen HELEN, the HERMIT OF THE SOUTHERN

MARCH, LETITIA Ketterley, King LUNE, MABEL Kirke, the High King PETER, Mr. and Mrs. PEVENSIE, and the unnamed KNIGHT who starves himself to keep Shasta alive. These characters have in common their honesty and care for others. We can assume that all these are summoned to the GREAT REUNION in LB.

The foolish and often wicked adults share a total contempt for things childish (see AHOSHTA, ALBERTA and HAROLD Scrubb, ARSHEESH, GLOZELLE, GUMPAS, MIRAZ, Miss PRIZZLE, PRUNAPRISMIA, Pug, RABADASH, SOPESPIAN, and the TISROC. In fact, it is a dead giveaway of wickedness in the *Chronicles* for an adult character to identify things children hold dear as "fairy tales" (PC 4) or "old wives' tales" (PC 9) and they usually pay for their nonbelief in the end. Miraz does not believe in the old STORIES and instructs Caspian never to talk or think about them; eventually (and almost as a direct result) he loses his life and his kingdom. Digory recognizes right away that Uncle ANDREW is an EVIL* MAGICIAN, and he knows from stories that magicians always come to a bad end. Lewis implies that if Andrew had believed these "old wives' tales" (MN 2), he might never have gotten started in the business.

ADVENTURE — In the *Chronicles,* adventure is a way of speaking about life at its highest intensity. There is no honorable turning away from the adventure, for it is only in leaving the known for the unknown that HONOR may truly be found. The meaning of adventure is perhaps made most clear to the two characters who have hardly an inkling of its meaning: JILL and EUSTACE, the two victims of modern SCHOOLS at EXPERIMENT HOUSE. Their IMAGINATIONS are unprepared to deal with the adventures they meet in NARNIA, but it is precisely in taking adventures as they come that they learn what adventure is (VDT 12; SC 13; LB 2, 9). Queen SUSAN is the first to use the word "adventure" (LWW 16), and she notes that people who have had similar adventures share a special speech and look.

[VDT SPOILERS] REEPICHEEP is perhaps the greatest adventurer in all the *Chronicles.* He is the first to leap into adventure and considers turning away to be COWARDICE. He proclaims the entire purpose of the *DAWN TREADER*'s voyage to be the search for honor and adventure and takes his own final adventure, alone and valiant, into the Utter East. His special quality consists of having no HOPES or FEARS, so that he is able to see the voyage into the darkness as pure adventure. Alone of all the crew he is untouched by the DARK ISLAND's horror (12). But Lord RHOOP and the Dawn Treaders enter the darkness, bringing their hopes and fears with them: They do not *take the adventure;* rather, the adventure *takes them.* Lost in their illusions, they almost lose their hope and their reason. Rhoop, indeed, is restored only by ASLAN's gift of DREAMless SLEEP. But Reepicheep reminds CASPIAN that the KING "shall not please himself with adventures as if he were a private person" (13), thus reminding Caspian of his VOCATION. Caspian himself appeals to the crew's sense of adventure when he asks them to continue on to the east (14).

In SC, RILIAN underscores Reepicheep's belief that adventure is not to be taken lightly. When the CHILDREN and PUDDLEGLUM are trying to guess the meaning of the fireworks, he says, "when once a man is launched on such an adventure as this, he must bid farewell to hopes and fears, otherwise death or deliverance will both come too late to save his honor and his reason" (SC 13). Rilian gives adventure another meaning. His FAITH in Aslan gives him confidence to say, "let us descend into the City and *take the adventure* that is sent to us" (my emphasis).

[LB SPOILERS] This acceptance of the necessity of adventure is echoed three times in LB (2, 9, and 12): once by TIRIAN, after JEWEL repeats that Aslan is not a tame lion; once again by Jewel himself, when he announces they should return to STABLE Hill; and again by Tirian when he accepts the fact that Jill and Eustace will stay with him for the inevitable LAST BATTLE. Just as HHB is an adventure for SHASTA and ARAVIS and BREE and HWIN; and as the adventure of exploring the attic leads DIGORY and Polly to greater adventures in Narnia; and as Narnia is an adventure for all of the SEVEN FRIENDS OF NARNIA, the *Chronicles of Narnia* are an adventure for the reader, which Lewis hoped they would keep with them for the rest of their lives.

AGING AND DISABILITY — In ASLAN'S COUNTRY those who have grown old become young and lose their gray hair and wrinkles, and the very young mature only to the flower of their manhood and womanhood.[1] This is beautifully illustrated by CASPIAN's transformation from DEATH to life at the end of SC (16), as well as by Lady POLLY's, Lord DIGORY's, and ERLIAN's in LB (12 and 16). King EDMUND's knee, damaged in a rugby match, is healed, and all are restored to perfect health, attire, and cleanliness (LB 13). Lewis, a man of his times and, yes, trying to see things through the eyes of CHILDREN, is still somewhat guilty of ageism, maybe because he himself was afraid of aging. For example, he has Jill say, "I'd rather be killed fighting for NARNIA than grow old and stupid at home and perhaps go about in a bath-chair and then die in the end just the same" (LB 9).

[AUTOBIOGRAPHICAL ALLUSION(S); YOUTH.]

[1]According to Christian theology, this age was thirty-three, since that was how old Jesus was at the end of his earthly life. See, for instance, St. Thomas Aquinas, *Sermons on the Apostles' Creed,* Article 11.

AHOSHTA TARKAAN — The TISROC of CALORMEN'S grand vizier (HHB 7 and 8), and the man to whom Aravis has been promised in marriage (HHB 3). Originally of the lowest class in Calormen, he has schemed and flattered his way into the highest circles of power. He owns three palaces, and an especially expensive one at the lake at Ilkeen. He is described as a "little, hump-backed, wizened old man" (7), and he is most often seen in a pose of prostration at the feet of the Tisroc. In Aravis's perception, he is a "hideous grovelling slave" (9) who pretends to be OBEDIENT but is really trying to manipulate the Tisroc for his own ends.

ALAMBIL — In Narnian ASTRONOMY, one of two planets in the night sky (the other is TARVA); it is surnamed "The Lady of Peace" by GLENSTORM the CENTAUR (PC 4 and 6).

ALBATROSS — A large sea-bird (VDT 12), and a symbol of good luck to sailors, who believed that to shoot one is to court bad luck.[1]

[ASLAN'S VOICE; DREAM(S); LITERARY ALLUSION(S); PROVIDENCE.]

[1]See the poem "The Rime of the Ancient Mariner," by Samuel Taylor Coleridge.

ALBERTA SCRUBB — The mother of EUSTACE and the aunt of EDMUND and LUCY. The CHILDREN dread spending their summer holiday at "Aunt Alberta's home," where even the one picture they like is banished to a small back room. She is a nonconformist, a vegetarian, teetotaler, and nonsmoker; both she and Harold wear "a special kind of underclothes" (VDT 1).[1] She is apparently something of a feminist, as indicated by Eustace's remark that she would consider CASPIAN's giving Lucy his cabin because she is a girl to be demeaning to girls (2). When the children return from the experience of the *DAWN TREADER* and remark how changed for the better Eustace is, Alberta only finds him commonplace and boring and blames his changed demeanor on the influence of "those Pevensie children" (16).

[ADULT(S); SEXISM.]

[1]These six traits—nonconformist, vegetarian, teetotaler, nonsmoker, wearer of "a special kind of underclothes," and feminist—would classify her and her husband and son not as Latter-day Saints but as members of one of the many back-to-nature groups of the early to mid-twentieth century.

ANDREW KETTERLEY — The mad MAGICIAN of MN [SPOILERS]. Andrew is DIGORY Kirke's uncle, the older brother of Digory's mother, MABEL Kirke, and resident in his sister LETITIA Ketterley's London house, where he lives on the top floor.[1] He is a pale shadow of his Narnian counterpart Jadis,[2] but through him Lewis provides a frightening suggestion of the destruction that can occur in our world if power is given over to the hands of immoral experimenters.[3] He is only out for himself: He has no scruples

As Mr. Tumnus, the Faun, serves tea, Lucy looks over the curious books on the shelves. (LWW 2)

about sending an unwitting Polly into an unknown world—one into which he does not himself dare to venture; he shows the depths of his despicableness when he silences Digory's protests by suggesting that any further noise might frighten his seriously ill mother to DEATH; and he thinks nothing about his godmother, Mrs. LEFAY, and his deathbed promise not to open her ATLANTEAN box. He shows himself to be the very antithesis of STOCK RESPONSES to human values at every step of the conversation he has with his nephew in Chapter 2. Rules are for CHILDREN, servants, women, and ordinary people—not for "geniuses" such as himself.[4] He has toiled his entire life to learn magic, at great cost to himself, so that he can open the forbidden box—and all it contains is dust (Lewis expects his readers to agree that this is the natural reward for a life devoted to magic). He attributes Digory's "warped" moral sense to his having been "brought up among women . . . on old wives' tales" (MN 2). In their experimenting with the RINGS, Digory and Polly discover that Uncle Andrew does not really know how the rings work; like most magicians, Lewis says, "Uncle Andrew . . . was working with things he did not really understand" (MN 3).

Andrew's extreme selfishness is well expressed in his inability to understand anything that doesn't directly relate to his own needs. He and the newly created TALKING BEASTS watch one another, the ANIMALS out of CURIOSITY and Andrew out of FEAR. He sees the animals only as potential threats and cannot understand their intelligent speech. For their part, they can't figure out whether he is animal, vegetable, or mineral, and their ensuing conversation about Andrew's nature is one of the funniest in all of the *Chronicles.* The miserable Andrew is eventually released by ASLAN from the cage in which the animals have kept him, and he spends the rest of his days living at the Kirkes' country house.

[GOLDEN TREE; HOUSE OF PROFESSOR DIGORY KIRKE; HUMOR; JACKDAW; RIGHT AND WRONG; SILVER TREE.]

[1]Polly's suggestion that Andrew is keeping a mad wife upstairs is a LITERARY ALLUSION to Rochester's action in *Jane Eyre,* by Charlote Brontë.

[2]See WHITE WITCH for a discussion of the parallels between Andrew and Jadis.

[3]Lewis thought a lot about the abuse of TECHNOLOGY, especially after nuclear weapons were developed. He would have the same misgivings about genetically modified food, cloning, and stem-cell research.

[4]This appeal is similar to that of the Grand Inquisitor in Dostoevsky's *The Brothers Karamazov.*

ANIMAL(S) — Animals of all sorts play a large part in the *Chronicles* and are present in all the books. Two types of animals inhabit NARNIA: DUMB BEASTS and TALKING BEASTS.[1] Although most animals are good and helpful, several are outstandingly bad. Thus giant bats are in the witch's army, and wolves are present for ASLAN's sacrifice. FENRIS ULF (Maugrim) is one of the witch's chief lieutenants. Apes assist in binding Aslan, and SHIFT is perhaps the most despicable animal of all.

Each animal acts according to its stereotype. Moles dig the apple orchard at Cair Paravel; Mr. BEAVER builds Beaversdam; HORSES carry smaller creatures into battle with the witch; GLIMFEATHER is a wise owl. Distinct from this, Lewis uses animals as hieroglyphs, or "pictures," of certain human attributes; REEPICHEEP, for instance, is a hieroglyph of COURAGE.[2]

According to TRUFFLEHUNTER, beasts (unlike humans and DWARFS) "hold on"—they do not change and they do not forget (PC 5). For this reason he is able to recall that Narnia was only "right" when a SON OF ADAM sat on the throne.

Finally, the regard in which animals are held is a good barometer of moral health. In PC, when Narnia is under the rule of the TELMARINES and the old STORIES are almost forgotten, it is observed that many more beasts are dumb than was the case in the GOLDEN AGE. Uncle ANDREW uses guinea pigs in his experiment with the MAGIC dust, and—because he owns them[3]—is not troubled that he has to kill them. And the fact that the GIANTS of HARFANG would eat Talking Stag—not to mention Men and MARSH-WIGGLES—shows that they are far from being good giants (SC 8).

[1]Animals from various world MYTHOLOGIES (such as the KRAKEN, the PHOENIX, and the UNICORN) are also present in Narnia; however, they seem to be a variety distinct from either dumb beasts or Talking Beasts.

[2]For a fuller discussion of Lewis's use of hieroglyphs, see TALKING BEASTS.

[3]As exemplified by the title of *The Horse and His Boy,* Lewis does not—at least in Narnia—recognize ownership of animals. In fact, they are presented as another sort of people. Hwin is spoken of as a gentle person (HHB 9), and the HERMIT calls the horses and goats his cousins (HHB 10). See also the unnamed hedgehog in HHB (12), "a small prickly person."

ANNE FEATHERSTONE — A schoolmate of Lucy Pevensie (VDT 10). Lucy is not fond of Anne, and Anne is jealous of Lucy's relationship with Marjorie Preston. Lucy magically invades Anne's privacy so she can overhear a conversation in which Anne shames Marjorie into pretending she does not care for Lucy at all, an act that hurts Lucy's feelings and leads to her vanity.

ANRADIN — A Tarkaan, remarkable for his scrupulously kept crimson beard. He visits the hut of Arsheesh and, in an act typical of an insensitive Calormene overlord, offers to buy his "son" Shasta (HHB 1). He is the master of Bree and thus by inference a veteran of the battle at Zalindreh. Later he is part of Rabadash's insurgents and a participant in the battle of Anvard (13). His fate is unknown.

ANVARD — The name of the castle of the king of Archenland in HHB (8, 12, 13, and 15). Very old, it is built of warm reddish brown stone and sits amid green lawns with a high, wooded ridge in back. It has many towers, but no moat. In his plot to take over Archenland and Narnia, Rabadash intends to hold Anvard and gather his forces there. It is the site of the battle of Anvard.

ARAVIR — In Narnian astronomy, the morning star (PC 11), most likely a planet like earth's Venus.

ARAVIS — A Calormene noblewoman, in her early teens, who is one of the four main characters in HHB (the others are Shasta, Bree, and Hwin). [SPOILERS] The story details her flight from the cruel, stifling world of southern Calormen to freedom in the north, specifically ARCHENLAND. More important, HHB is the chronicle of her transformation from arrogance and self-centeredness into an example of true Narnian nobility, that is, the exercise of humble and compassionate leadership. She becomes queen of Archenland (the wife of King Cor) and the mother of King RAM the Great. She is last seen in the assembly of famous Narnians in the GREAT REUNION in ASLAN'S COUNTRY (LB 16).

[AUTOBIOGRAPHICAL ALLUSION(S); HORSE(S), HORSEMANSHIP; SEXISM.]

ARCHENLAND — In Narnian GEOGRAPHY, the smaller and more southern of the two northern kingdoms, the other being NARNIA. It is bordered to the south by the Southern MARCHES (where the HERMIT lives), and to the north by the Northern Mountains (which are, of course, Narnia's Southern Mountains). Archenland is a lovely country of gentle hills, snow-clad, blue-peaked mountains, and narrow gorges. The slopes are covered with pine, and all varieties of TREES grow in the parklike portions of the country. Archenland and Narnia have been friends since before MEMORY, and the High King PETER intends to make CORIN a KNIGHT at CAIR PARAVEL (PC 5). The WINE of Archenland is so strong that it must be mixed with water before being drunk (VDT 6). In MN 11, ASLAN foretells that some of the descendants of King FRANK and Queen HELEN will be KINGS of Archenland.

ARGOZ — In VDT 2, 9, 13, and 14, one of the SEVEN NOBLE LORDS, and one of the four TELMARINE visitors to the land of the DUFFERS in the years 2299–2300 N.Y. At ASLAN'S TABLE he is one of the THREE SLEEPERS. When Lord RHOOP is rescued from the DARK ISLAND, he is seated beside Argoz to begin his rest.

ARSHEESH — A poor fisherman who lives in the far south of CALORMEN on a little creek of the sea[1] with his "son" SHASTA. The awkward expression "with him there lived a boy who called him father" is indicative of something irregular in their relationship. SILENT and distant, the ADULT Arsheesh is constantly finding fault with the boy and sometimes beats him. His practical mind and limited vision make him unable to answer Shasta's questions about the north, the land of freedom. As a result, the boy becomes wary and uncommunicative toward grown-ups in general. There may be an intended similarity between the name "Arsheesh" and the word that best describes him, *harsh* (HHB I, 5, and II).

[1]Note Lewis's use of the British meaning of *creek* as a small inlet or bay that is narrower and extends farther inland than a cove. Confusion with the American meaning has caused some misunderstanding of the picture Lewis has in his mind of GLASSWATER CREEK.

ASLAN — The Lion King of the land of Narnia and of all its creatures, the son of the EMPEROR-BEYOND-THE-SEA, true beast and the KING of beasts, the highest king over all high kings, and the as-yet-unrecognized good and compassionate Lord of all, beginning with the CHILDREN from ENGLAND. To hear his name is an experience of the NUMINOUS for all who are destined to live in his country, but for those who are for a time or forever under the spell of EVIL*

The stare of the White Witch makes Edmund uncomfortable: She is tall and beautiful, proud and stern; her red lips the only spot of color in a face pale as ice. (LWW 3)

MAGIC his name is filled only with horror. To see his beautiful face sustains one all one's days, and the recognition of that face with love and awe at the end of TIME opens out onto an eternity of JOY. To be addressed by him as "dear heart" or "little one" or by name is a lasting, cherished blessing; to be rebuked by him is an everlasting shame. The person he praises with an earthshaking "Well done" remains forever favored; the person he blames or punishes is humbled in the hope of an enduring change of heart. Though he is wild—that is, all-powerful and free—he delights to be at the center of the DANCE of those whom he has made; he welcomes the help of others, both beast and human, to accomplish his plans; and he is the very often unnoticed storyteller behind every person's STORY, guarding the PRIVACY of each, keeping FAITH with all. LUCY and CASPIAN and REEPICHEEP seem to be the English woman and the Narnian man and the Narnian TALKING BEAST most beloved of Aslan in the *Chronicles* only because more is told of their stories.

Physical Appearance — No one has ever seen anything more terrible or beautiful than Aslan; in this respect, he is the perfect example of the majestic, the glorious, and the numinous. He is towering in size, larger than a HORSE, as large as a young elephant, and always growing bigger with respect to the person who sees him; in this respect, he is the very figure of the GREATNESS OF GOD. In overall aspect, he is "so bright and real and strong" (SC 16) that all else pales in comparison; light seems to radiate from him. His coat is a "soft roughness of golden fur" (LWW 15), ranging in color from tawny gold to bright yellow. During Aslan's ecstatic romp with SUSAN and Lucy on the morning of his resurrection, his tail lashes back and forth in intense joy. His paws are beautifully velveted in FRIENDSHIP and terrible in battle; although heavy enough to make the earth shake, he walks noiselessly, as do all his feline relatives. His legs, haunches, shoulders, chest, and back are powerfully muscled. His shaggy mane is a beautiful sea of rich, silky, golden fur, scented with a solemn, strengthening perfume. He uses his long whiskers to prove his lion-ness to the doubting BREE. His golden face reveals his regal personality in all of its emotions. This is especially true of his "great, royal, solemn, overwhelming eyes" (LWW 12), which reflect the full range of his FEELINGS, from happiness and mirth to scorn and anger. If it weren't for the calming quality of ASLAN'S VOICE, no one could stand in his awesomely beautiful presence.

ASLAN'S BREATH — In the Judaeo-Christian tradition, the Spirit of God is the creative, strengthening, renewing power of God, symbolized by the breath of life, the wind, fragrant oil, water, and fire. In the Christian doctrine of the TRINITY, the Holy Spirit is the third Person, the principle of sanctification; in the tradition of Western Christianity, he comes forth from the Father and the Son, as in the gospel scene where Jesus, after his resurrection, breathes on his apostles and says, "Receive the Holy Spirit."[1] In the *Chronicles* Lewis reflects his profound assimilation of this Judaeo-Christian tradition in the rich symbolism with which he surrounds ASLAN.

The most explicit Narnian reference to the Holy Spirit

is in the marvelous scene where the Large Voice speaks a threefold "Myself" in answer to SHASTA's question, "Who *are* you?" The third "Myself" is a nearly inaudible whisper that seems to come from everywhere, as if the leaves rustled with it.[2] Lewis here intends to give a sound-picture of the subtle, yet all-pervasive, activity of the Spirit of God.

Another allusion to the activity of the Spirit is found in FATHER CHRISTMAS's outfitting of PETER, SUSAN, and LUCY with their gifts from Aslan. They are given instruments of combat, surely an allusion to Ephesians 6:11–18, in which the virtues of truth, righteousness, peace, FAITH, salvation, and the word of God are the belt, breastplate, shoes, shield, helmet, and sword of the soldier.

Beyond these two allusions, Aslan's breath (with its sometimes emphasized fragrance) is the chief symbol of the Spirit's activity in the *Chronicles.* After his resurrection Aslan breathes on Susan to reassure her that he is not a GHOST; he breathes on the creatures turned to stone in the White Witch's castle and they return to life. In PC, he breathes on Susan to quiet her FEARS and twice on Lucy—once in a sheer outpouring of love, and a second time to give her strength to face her companions; EDMUND looks formidable to the Lords GLOZELLE and SOPESPIAN because, when Aslan and Edmund had met, the Lion had breathed on him "and a kind of greatness hung on him" (PC 13); and when Aslan's breath comes over the brave TELMARINE soldier, the man wears a happy but startled look, "as if he were trying to remember something" (PC 15)—he now has the COURAGE to walk through the DOOR. In VDT 12, Lucy is heartened by ASLAN'S VOICE, speaking through the ALBATROSS, and by the "delicious SMELL" of his breath. In SC 1 and 16, EUSTACE and JILL are blown into NARNIA from ASLAN'S COUNTRY and from Narnia back into his country by Aslan's gentle but powerful breath. In HHB 11, the Large Voice breathes on Shasta to assure the boy that it is not a ghost. In LB 15, in his last gesture before he leaves to make his final appearance at the GIANTS' stairway, Aslan breathes on EMETH, taking away his fear.

In MN, the allusions to the creative and spiritualizing activity of the Holy Spirit are very strong. After the creation of the STARS, "a light wind, very fresh, begins to stir" (8). This will remind the adult reader in the Judaeo-Christian tradition of the first verses of the book of Genesis, the first book of the Bible. That Lewis intends this allusion is clear from the next reference to the wind, in which he uses the definite article: "the light wind could now be ruffling the grass" (9). When Aslan confers the GIFT OF SPEECH on the chosen ANIMALS (9), his "long, warm breath . . . seemed to sway all the beasts as the wind sways a line of trees. . . . Then came a flash of fire (but it burnt nobody) either from the sky or from the Lion itself, and every drop of blood tingled in the CHILDREN's bodies." This passage is remarkable for the intense breath image and the addition of the fire image (from the first conferral of the Holy Spirit in the New Testament on Pentecost; see the Acts of the Apostles 2:3–4).[3]

One of the most striking symbols for the Spirit's activity in Narnia is the SWEET waters of the LAST SEA. These fortify all who drink them with the power to look directly at the SUN and they diminish the need for food and SLEEP. In Christian theology, the Spirit prepares the person to see the glory of God by transforming the person gradually into God's likeness.[4]

[Biblical allusion(s); Memory; Profession(s) of faith.]

[1]Gospel of John 20:22.

[2]HBB 11. Alert readers of Lewis will recall a similar, powerful image in his sermon "The Weight of Glory" in *The Weight of Glory,* ¶12.

[3]For the allusion to the tingling feeling in the children's blood, see Gift of speech.

[4]2 Corinthians 3:18.

ASLAN'S COUNTRY — The land that is home to Aslan and to all creatures who recognize Aslan with joy. It is a range of incredibly high yet snow-free mountains, bathed in late-spring/midsummer warm breezes and freshness, alive with the sounds of running water, waterfalls, and birdsong against a "background of immense silence" (SC 1) and covered with orchards of autumn-ripe fruit, forests of mighty trees, and flower-decked meadows. It is not connected with any created country, but every real country is connected with it, as peninsula to mainland. It lies beyond the edge and beyond the sun of every world and can be reached only by magic or through the door of a noble death. No one is any particular age there: All come into the full flower of their manhood or womanhood (SC 16). It is also known as Aslan's Land and as Aslan's Mountain.

Aslan's country is not mentioned in LWW, in HHB, or in MN, and its single mention in PC 4 is in Doctor Cornelius's story to the young Prince Caspian. In VDT, it is the object of Reepicheep's quest and part of Edmund's profession of faith. The sight of it at the World's End fades for Edmund, Eustace, and Lucy as Narnia's sun rises. SC's transitions between England and Narnia

Susan, Lucy, and Edmund help Mrs. Beaver prepare dinner as Peter and Mr. Beaver come in with the main course, fresh-caught trout. (LWW 7)

take place through Aslan's country; it is in this book that we discover that the air there is clearer than in Narnia or in other worlds. So, too, minds are clearer in Aslan's country—down in Narnia and other countries the air is thicker and minds more confused and confusable.

The most significant observations about the nature of Aslan's country are made in LB. JEWEL'S HOPE is that the DOOR of the STABLE may lead to Aslan's country and to ASLAN'S TABLE. TIRIAN, having ushered RISHDA into the arms of TASH through the door, discovers himself refreshed, cooled, cleaned, and dressed for a feast. The pleasures of eating the wonderful fruit there not only are not wrong (and therefore forbidden) but are entirely RIGHT and lawful and encouraged. And when Lucy, PETER, and Edmund have their power of vision magnified immeasurably, they see that the mountains of the Utter East ring Narnia round to become the WESTERN WILD, at the heart of which is the GARDEN in which they are standing. In addition, they see the real England and their parents, Mr. and Mrs. PEVENSIE, waving to them. TUMNUS explains that the heart of each good culture survives because it has always been a part of Aslan's country. Finally, all are welcome to go further up into these mountains and further in this land—the real ADVENTURES have only begun.

ASLAN'S HOW — In LWW 12, 14, and 15, it is the round, high, open hilltop south of the GREAT RIVER and west of the River Rush on the edge of the GREAT WOODS. From its brow, a person can look down upon all the forests of NARNIA and the Eastern Sea in the distance. Its other name is the Hill of the STONE TABLE because this ancient place of sacrifice is in the very center of the hilltop. Lewis chooses this sacred site as the place where the three English children, PETER, SUSAN, and LUCY, first[1] meet ASLAN in a magnificent heraldic tableau.

In PC 7 and 11, whose story takes place thirteen hundred Narnian years after the events of LWW, this hill has gotten the name Aslan's How (from the Old Norse and Old Teutonic name for mound or cairn). A second huge round hill has been raised by ancient Narnians over the broken Stone Table, which is now referred to simply as "the Stone." The mound is hollowed into mazelike tunnels, galleries, and caves, all lined and roofed with smooth stones, and carved with ancient writing, snaking patterns, and—everywhere—stone reliefs of Aslan (like Celtic shrines and burial places). At this point, the place is also known as the Great Mound, or simply the Mound. Prince Caspian's army makes its headquarters here during the WAR OF DELIVERANCE. And though heretofore perhaps the most hallowed place in Narnia, it is not mentioned again in the *Chronicles,* except once, in passing, in HHB 12.

[GEOGRAPHY, NARNIAN.]

[1]Since LWW was the first of the *Chronicles* to be written, this fact is especially significant. See *Advice to an Intelligent Reader from an Intelligent Reader.*

ASLAN'S NAME — It is the tradition among the CENTAURS that ASLAN[1] has nine names (SC 16); but we are given only four, and these may not be among the nine the Centaurs know. They are Aslan, the great Lion, the son of the EMPEROR-BEYOND-THE-SEA, and the King above all High Kings.

In addition, before he is killed on the STONE TABLE, Aslan is mocked by the White Witch and her minions, who call him "the great fool" and "the great CAT"—the first may be an allusion to "the Fool in Christ," a special kind of saint in the Orthodox tradition.[2]

True to his own principles about good writing,[3] Lewis, in his first several uses of the name of Aslan, does not throw descriptive adjectives and adverbs at his readers but tries to elicit from their IMAGINATIONS and MEMORIES the experience of the "enormous meaning" some names have in DREAMS. Everyone has experienced a dream turn into a nightmare or a dream become so beautiful that it becomes a lifelong memory and kindles the desire always to live in that beauty. It is *this* NUMINOUS terror and *this* delight that the name of Aslan evokes in each of the Pevensie CHILDREN (though for Edmund, given his addiction to the black MAGIC of the TURKISH DELIGHT, it occasions only horror until he undergoes his conversion).

In VDT, the power of Aslan's name, especially PRAYER in his name, is highlighted. CASPIAN undertakes the voyage into the darkness surrounding the DARK ISLAND only after invoking the name of Aslan. It is in answer to LUCY's prayer to Aslan that the ALBATROSS comes to lead the *DAWN TREADER* out of that same darkness. Caspian demonstrates the urgency of his desire to help the THREE SLEEPERS and to express his growing love for the STAR'S DAUGHTER by asking her in Aslan's name how to disenchant them. Then, in a closing scene that suggests a major reason why Lewis wrote the *Chronicles,* Aslan reveals to Lucy that he has another name in our world and that the reason the children were brought to NARNIA was to know him by his Narnian name in order to be able to recognize him by his earthly name.[4]

Perhaps still under the spell of the ending to VDT, Lewis intensifies his own fictional exploration of the "theology" of Aslan's name in SC. JILL and EUSTACE, having refused to force their way into Narnia by the use of magic, have time enough to invoke Aslan's name only three times before they are interrupted by members of the GANG. This is quite sufficient, however, for their TRANSITION into ASLAN'S COUNTRY. And, as part of her QUEST (her temporary VOCATION), Jill is given four SIGNS, the last of which is that she will know who Prince RILIAN is because he will be the first person to ask her something in Aslan's name. MN contains the most exalted insight about the meaning of Aslan's name. In the great scene (10) in which the GIFT OF SPEECH and thus reasoning is conferred upon the ANIMALS, Aslan's first use[5] of the name "Narnia" in his command to the land and its creatures to awaken is reciprocated by the very first word *they* speak: his name. This is an awesome hint at and a beautiful picture of what Lewis believed to be the deepest meaning of creation itself: Whatever the full meaning of God's desire to create anything outside himself, humankind realizes its vocation when it becomes the priest for dumb creation and speaks the name of God back to him.[6]

[1]In Turkish, *aslan* is the word for "lion." Lewis found the name in the notes to Edward William Lane's *Arabian Nights.* He pronounced the name "ASS-lan" (see *Letters to Children,* 29). *As* is also an old Scandinavian word meaning "god," as in *Asgard,* the home of the gods in Norse MYTHOLOGY.

[2]Kallistos Ware, *The Orthodox Way* (rev. ed.: Crestwood, NY: St. Vladimir's Seminary Press, 1995), 95. This reference is supplied by Joshua Falconer.

[3]Lewis received many letters from would-be writers, even children, asking his counsel on the writing craft; and true to his generous nature, he answered every one. In almost every answer, he included the following advice:

> Never use adjectives or adverbs which are mere appeals to the reader to feel as you want him to feel. He won't do it just because you ask him: you've got to *make* him. No use telling us a battle was "exciting." If you succeed in exciting us the adjective will be unnecessary; if you don't, it will be useless. Don't tell us the jewels had an "emotional" glitter; make us feel the emotion. I can hardly tell you how important this is. (Letters 1988, 2 September 1957)

[4]See *Advice to an Intelligent Reader from an Intelligent Reader* and *Letters to Children*, 32.

[5]These "first usages" are first only with respect to the internal chronology of the *Chronicles.* This entry, as well as this entire *Pocket Companion to Narnia,* tries to deal with themes in the order of the essential completion of the books, as explained in the *Advice to an Intelligent Reader from an Intelligent Reader,* 11–14.

[6]The most striking picture of this truth comes in the song of nature (modeled on Psalm 110) in Lewis's *The Great Divorce* (Chapter XI), which concludes: "Master, your Master has appointed you for ever: to be our King of Justice and our high Priest." The first "master" is the man who has overcome his lust; the second "master" is God; and the lust, converted, is now a magnificent HORSE.

ASLAN'S TABLE — Located on RAMANDU'S ISLAND, the table is surrounded by richly carved stone chairs with silk cushions (VDT 13). It runs the length of a clearing, which is paved with smooth stones and surrounded by large pillars, but open to the sky. A crimson cloth covers the table nearly to the ground, and the STONE KNIFE rests upon it at all times, almost as if enshrined. When the DAUGHTER OF RAMANDU (who first calls it "ASLAN's Table") sets her candlesticks upon it, all the gold and silver accoutrements glow richly in the light. By Aslan's decree, each evening at sunset the table is MAGICALLY laden with a banquet far grander than any ever set in the GOLDEN AGE OF NARNIA. If this food is not eaten by travelers before daybreak, it is consumed by the birds of morning, and so the table remains empty until the next sunset. A place of magical renewal, the table—with its medieval setting and grail-like knife—suggests the spirituality of the ancient Arthurian legends; and being Aslan's table, it also suggests the eternal refreshment of the Eucharist as the heavenly banquet.

[PROFESSION(S) OF FAITH.]

ASLAN'S VOICE — One of the greatest advantages in the choice of a talking lion to be the hero of the *Chronicles* is the addition of a whole repertory of roarings, growlings, and purrings to the ordinary range of human speech expression. Beyond understanding what ASLAN actually says, the reader's chief way of gauging the Lion's thoughts and FEELINGS is through the quality of his voice. The fact that DIGORY'S TREE in London does not grow up to be fully MAGICAL since it is far away from the sound of Aslan's voice (MN 15) is an instance of the Lion's voice being an integral part of the atmosphere of Narnia. It is "deeper, wilder, and stronger" than a man's voice, "a sort of heavy, golden voice" (SC 2).

Aslan's roar is reserved for celebrating his chief exultations (most notably his resurrection in LWW 14), expressing his greatest displeasure (for example, at the WHITE

Susan and Lucy watch in awe as Aslan's gentle face grows terrible to behold and the very trees bend before the power of his roar. (LWW 16)

WITCH's suggestion that he might go back on his promise in LWW 13), announcing his victory (as when he kills the witch in LWW and when he returns to Narnia in PC 11), and issuing his most momentous commands (such as calling on FATHER TIME to begin the END OF NARNIA in LB 13).

Aslan's growl is usually probing, more tentative, as when he compels Digory to take responsibility for his actions (MN 11) or when he attempts to draw the renegade DWARFS into the JOYS of ASLAN'S COUNTRY (LB 13). He also growls at his loved ones when they would misunderstand him or his ways (PC 15). But it is only with his loved ones that he also purrs, one of his rarer expressions in the *Chronicles;* Aslan purrs his love for LUCY with a low, EARTHQUAKElike sound (VDT 12) and purrs at the sound of SUSAN'S HORN in LWW (12).

In a performance reminiscent of Tolkien's Music of the Ainur,[1] Aslan sings Narnia and its creatures into existence (MN 8–10). Beginning wordlessly and nearly tunelessly, the Lion's song grows in strength and glory until the SUN rises, becomes gentle and lilting for the growing of grasses and trees, and, after a wildly passionate section, lapses into a solemn SILENCE. But Aslan's "deepest, wildest voice" is used only for bestowing the GIFT OF SPEECH upon the chosen Narnian ANIMALS.

Aslan's shouts are set aside for approvals and encouragements; note his earthshaking "Well done" to Digory (MN 14) and his invitation to the SEVEN FRIENDS OF NARNIA and the blessed Narnians: "Come further in! Come further up!" (LB 14). In contrast, his somberest moods are reflected in his "dull voice," as on the evening of his DEATH (LWW 14).

Finally, his voice alone communicates the profoundest mystery of the TRINITY to SHASTA (HHB 11). Aslan is his voice, the "Large Voice," to the boy surrounded in the night fog; and the threefold distinction of voices (the first, very deep, low, and earthshaking; the second, "loud and clear and gay"; and the third, the almost inaudible but all-encompassing whisper) is one of the singular achievements of Lewis as a remythologizer.

[1]J. R. R. Tolkien, *The Silmarillion* (Boston: Houghton Mifflin Co., 1977; London: George Allen & Unwin, 1977), 15–17.

ASTROLABE — See CORIAKIN.

ASTRONOMY, NARNIAN — In the *Chronicles,* Lewis assumes a medieval worldview of a NARNIA-centered universe.[1] Narnia itself is flat—CASPIAN has heard STORIES about worlds that are round, like balls, but he never believed they were true (VDT 15). The planets are great lords and ladies; the STARS are wondrous, silvery beings. As befits a medieval universe, the portents of astrology are taken seriously, and CENTAURS are especially good at reading these omens in the heavens. In many ways, the Narnian planetary system is very much like our own: The day is twenty-four hours long (Orruns immediately recognizes the reference when EUSTACE says it's after 10 o'clock [SC 16]); the SUN rises in the east and sets in the west and has a companion moon. The Narnian moon is larger than ours. After it has set, ARAVIR, the morning star, gleams like a little moon in the east. In HHB, the moon is said to be behind Aslan as he bounds on the scene between Shasta and the desert. This *suggests* that CALORMEN is far to the south of

Narnia and ARCHENLAND. Many constellations can be seen in the night sky, among them the summer constellations of the Ship, the Hammer, and the LEOPARD. There is apparently some scientific study of the night sky, as a number of astronomical instruments are found in the HOUSE OF THE MAGICIAN (VDT 11). Lewis himself owned a telescope (see AUTOBIOGRAPHICAL ALLUSIONS).

[1]See EMPEROR-BEYOND-THE-SEA, n. 2.

ATLANTEAN, ATLANTIS — ANDREW Ketterley surmises that the box of his godmother, Mrs. LEFAY, is Atlantean (MN 2), that is, a remnant of one of the oldest civilizations on earth. Whether or not Atlantis really existed has never been proven, but traditionally it is thought to have been an island west of the Strait of Gibraltar that long ago vanished into the sea.

AUTOBIOGRAPHICAL ALLUSION(S) — C. S. Lewis's childhood was marked by the DEATH of his mother when he was nine, and his troubled relationship with his father complicated his YOUTH. Only in the fleeting scenes of DIGORY's mother's recovery of her health and her playing with POLLY and Digory and of the meeting of TIRIAN with his father, ERLIAN, does the reader catch a glimpse of the kind of healthy relationship of parents, indeed ADULTS, and CHILDREN that is so notable in the works of Lewis's mentor, George MacDonald. Thus ARAVIS's father is heartless; SHASTA's "father" is sullen and mean (the boy has lost his mother and does not know his real father); CASPIAN is an orphan and his father figure, MIRAZ, is a tyrant; Digory's father is in India; and Mr. PEVENSIE is frequently absent.

Digory's desperate desire to help his seriously ill mother reflects Lewis's crisis over his own mother's illness, during which he experienced the appeal of MAGIC and the failure of petitionary PRAYER, the immediate cause of his abandoning any FAITH in an all-powerful, all-good God. Lewis revisited this terrible time in his life at the time of his wife's death and wrestled with his belief in an all-good God, the thought of SUICIDE briefly flitting across his mind as he worried that humanity is like rats in God's trap or in God's laboratory, with God enjoying the PAIN He causes us.

Other patterns that emerge in the stories from Lewis's life are his love of the sea, of books, of DOMESTICITY, and of nature and walking and his loathing of INSECTS and SCHOOLS. Lewis also loved CATS and DOGS. EUSTACE's fall over the cliff and JILL's vertigo in SC reflect Lewis's own FEAR of cliffs. His delight in bathing is conveyed by Eustace's bath at ASLAN's "hands" in Chapter 7 of VDT. As he grew older, it seems he became afraid of AGING AND DISABILITY.

The rigorous education Lewis received from his tutor, W. T. Kirkpatrick, is echoed in Dr. CORNELIUS's relationship with Caspian X and in Professor Digory Kirke's interaction with the Pevensie children, especially when he invokes PLATO again and again.

Three London girls came to live at Lewis's home, The Kilns, at the beginning of World War II to escape the anticipated bombing of London: Margaret, Mary, and Katherine (the last rumored to have later been killed in the blitz with her parents). Margaret wrote me the following letter on February 11, 1977:

I was evacuated with my fellow students of the Convent of the Sacred Heart, Hammersmith, and was fortunate to be billeted (with two other girls) with Mrs. Moore at The Kilns, Headington, Oxford . . . ¶ You may like to know something of the house; it was fairly large, stood in its own grounds which incorporated large lawns, flower beds, a tennis court, natural lake, copse and woodland leading up to Shotover, a large kitchen garden, a bungalow which we girls were allowed to use for our studies, and a summer house. ¶ Mrs. Moore, a widow, lost her only son in the 1914–1918 war. C. S. Lewis and his brother, Major Lewis, were great friends of her son, and as their parents were dead Mrs. Moore adopted them—whether legally or not I do not know. ¶ I saw little of Major Lewis as he was in the army, so the household consisted of Mrs. Moore, her daughter Kitty, C. S. Lewis, a cook, a parlour maid, and a gardener. ¶ It was obvious that Mrs. Moore was devoted to Lewis; she was over-protective and I felt at the time she still thought of him as a small boy; she called him "Boyboys" and he called her "Mintons." ¶ I shall always remember one warm, sunny spring day; we were all at lunch in the summer house—a distance of 20–30 yds. from the house—when it grew overcast and before the meal was finished it started to rain quite heavily. Lunch over, Mrs. Moore rang for the parlour maid to ask her to fetch an umbrella and galoshes for Mr. Lewis so he could return to the house and not get wet. ¶ Lewis, I am certain, was liked and respected by his students. Often at weekends 3 or 4 came to the house (always male students—Mrs. Moore would I am sure not have taken kindly to females) and he played tennis with them and went swimming or boating on the lake. We girls joined in these activities—being school girls Mrs. Moore considered it safe for us to be with "Boyboys." ¶ I was taking my School Certificate in that June and Lewis was a great help to me, always interested and willing to give advice. ¶ Mrs. Moore was very Victorian in her outlook and in her dress, and although I was 17 years old (the other 2 girls were younger) I was never allowed to have dinner with the family. We had supper which consisted every night of marie biscuits, an apple and a glass of milk. Without the help of Lewis and the cook we would have spent many a hungry night. ¶ The bedroom which we girls shared was above Lewis's study, which had a bay window with a flat roof. He used to pass food up to us and often helped us down so we could visit the kitchen where cook gave us food. Sometimes we climbed through the window of his study and listened to his records with him. On occasion he took us to the local fish and chip shop and we'd eat our secretive meal out of boxes on the way home. ¶ One May morning he invited us to the top of Magdalen Tower to hear the singing. We often had tea with him after school in his rooms at college. ¶ Once he took me to meet Masefield, and on another occasion I met Tolkien. It was then I heard Tolkien and Lewis discussing the "Lord of the Rings" and I feel looking back that the embryo of the Narnia series began to take shape. ¶ Lewis was a keen astronomer and had a telescope on the balcony of his bedroom. I was privileged to be shown many of the

> wonders of the universe. ¶ He was a wonderful story teller and would tell us tales as we sat in the garden or walked through the woods and over Shotover. ¶ He was unpretentious, a casual dresser preferring tweeds or grey flannels and sports jacket, usually carried a stout walking stick and always wore a deer-stalker hat. ¶ He seemed unconcerned with the war, his mind being filled with space, the heavens, literature and his church. ¶ I think he was disappointed we girls were Catholics; he asked us once or twice to go and hear him preach (he was a lay preacher) and in return he came with us to Mass once or twice. ¶ . . . He was a kind, sympathetic and very human man, never talking down to us school girls. I shall always consider it a great privilege to have known him.

One would have to say that Lewis's experience of being a professor in a country house in which children had taken refuge during wartime was a catalyst for the *Chronicles of Narnia.*

Lewis's particular affection for the children of some of his dear friends is the source of the DEDICATIONS of the *Chronicles.* Many of these dear friends were members of his writing group, the Inklings, who met twice a week from 1933 to 1963 during the academic year and included J. R. R. Tolkien, Warren Lewis, Hugo Dyson, Charles Williams, Dr. Robert Havard, Owen Barfield, Neville Coghill, and others.

The best book-length biography of C. S. Lewis for younger readers is *C. S. Lewis: Creator of Narnia,* by Elaine Murray Stone.[1]

[1]New York: Paulist, 2001; ISBN 0–8091–6672–0.

AUTUMN FEAST — Two Autumn Feasts are mentioned in the *Chronicles:* the feast of the "Gentle GIANTS" of HARFANG, in which JILL and EUSTACE are to *be* the feast (SC 6, 8, and 9); and the feast of the CALORMENES, at which the ass RABADASH, stepping into the temple of TASH, returns to human form (HHB 15). The autumn feast of Samain, which took place in November, was one of the four main feasts of the Celtic year in ancient Britain. The year began on this day for the Celts; much ritual, of which human sacrifice was quite probably a part, was associated with it. This was a time of great tension for the ancient Celts, in which the supernatural and material worlds drew close to each other as the earth moved into the darkness of winter.[1] Thus the trek of the three travelers across ETTINSMOOR, during which they meet the QUEEN OF UNDERLAND and the black KNIGHT, is quite appropriate; they do enter into the darkness of Underland, the spell of which is broken only by the DEATH of the queen.

[1]*New Larousse Encyclopedia of Mythology* (London: The Hamlyn Publishing Group Ltd., 1978), 236.

BACCHUS — In Roman MYTHOLOGY, the GOD of WINE and ecstasy[1] (in Greek mythology, he is named Dionysus), also known as Bromios, Bassareus, or the Ram. He is a youth dressed only in a faun-skin, and vine leaves wreathe his curly hair. His face is wild, and almost too pretty for a boy. EDMUND thinks he looks capable of doing absolutely anything. He is often accompanied by SILENUS and the MAENADS, with whom he dances the wild DANCE of plenty (LWW 2 and PC 11 and 14).

[1]The cry "Euan, euan, euoi-oi-oi-oi" is the traditional ecstatic utterance at the rites and feasting of Bacchus. "Euhan" or "Euan" is a Greek surname of Bacchus and *euhoi* or *euoi* is an interjection, a shout of joy heard throughout a bacchanal (PC 11 and 14.)

BANNER, STANDARD, CROWN, CORONET —

Mentioned throughout the *Chronicles,* these emblems of KINGS and queens and of CHIVALRIC ORDERS help lend a medieval flavor to the ADVENTURES.

Banners, Standards — The Narnian flag is a rampant lion (ASLAN, of course) on a green field. In LWW 12, the lion Lucy sees is red, as is the lion that graces the banner carried by Lord Peridan. However, the great banner that flies at half-mast above the castle at Cair Paravel in SC 16 bears a gold lion. Whether the color of the lion varied, or whether Lewis simply forgot what color the lion was, is a matter for speculation.

Crowns, Coronets — Narnian crowns are "light, delicate, beautifully shaped circlets," (MN 14), especially in contrast to modern European crowns. People actually look better when wearing Narnian crowns. DWARFS made the gold circlets worn by King FRANK and Queen HELEN, his covered with rubies and hers with emeralds (MN 14). In VDT 15, Lucy can tell that the group of SEA

PEOPLE she is seeing is lordly and noble, because they are wearing coronets.

BAR — Former lord chancellor of ARCHENLAND, he was dismissed by King LUNE for embezzling royal monies and spying for the TISROC (HHB 14). When the CENTAUR prophesies that Prince Cor will someday save Archenland from great danger, Bar kidnaps the boy and flees to CALORMEN. However, before he can reach his destination, his galleon is overtaken and Lord Bar is killed. Cor is renamed SHASTA by ARSHEESH.

BASTABLES — The famous fictional children created by Edith Nesbit (1858–1924), one of C. S. Lewis's favorite writers.[1] They are mentioned in the same context as SHERLOCK HOLMES, to give further support to Lewis's sense that fictional characters "really" exist in the world of fiction (MN 1). The Bastable children are the heroes of *The Story of the Treasure Seekers* (1899), *The Would-Be-Goods* (1901), and *The New Treasure Seekers* (1904). These were collected in 1928 as *The Bastable Children.* Lewis often recommended Nesbit's other cycle of stories about four children and their baby brother (Cyril, Robert, Anthea, Jane, and the Lamb), *The Five Children and It* (1902).

[1]Lewis had probably as a child read E. Nesbit's story "The Aunt and Amabel," in which Amabel finds her way into a MAGIC world via "Bigwardrobeinspareroom." See LITERARY ALLUSION(S) and WARDROBE.

BEAVER, MR. — A cordial, hardworking TALKING BEAST, proud of his dam-building skills and lovingly content with his domestic situation. He seems to be a hieroglyph of the sturdy, working-class Englishman. He befriends the four CHILDREN in the Narnian wood, proving to LUCY that he is trustworthy by showing her the handkerchief she had left with TUMNUS. He invites the children to dinner at his cozy home. There he introduces the children (and us, his readers) to some Narnian HISTORY and characters.

Mr. Beaver tells of the EVIL power of the WHITE WITCH and of the character of the great ASLAN. A creature of deep and simple FAITH, Mr. Beaver recites the three ancient prophecies that tell of Aslan's triumph and the end of the Hundred Years of Winter as well as foretelling that the coming of four human beings is to figure in the future of NARNIA. Later he makes a PROFESSION OF FAITH in Aslan as safe but not tame.

An astute judge of character, Mr. Beaver knows immediately where EDMUND has gone and why. He tells the children that only Aslan is capable of rescuing Edmund and overcoming the White Witch. Then, in keeping with this faith, he and his wife lead the group on the journey to escape the witch and to meet Aslan at the STONE TABLE. His loyalty to Aslan, always apparent, erupts into anger when he hears the witch hailed as Queen of Narnia and into anxiety and PRAYER as he holds Mrs. BEAVER's paw during Aslan's private talk with the witch. True to Aslan and those whom Aslan loves, he shares in the rewards and honors at the children's coronation and is among the blessed at the GREAT REUNION (LB 16).

BEAVER, MRS. — A kind old she-beaver devoted to her husband, Mr. BEAVER, she is happily busy at her sewing machine most of the time. She keeps a snug home, which is decorated in a sea motif. She seems to be a hieroglyph of DOMESTICITY. When the CHILDREN arrive, she greets them with a Simeon-like "To think that I should see this day." In speaking of ASLAN, she says that anyone who does not FEAR him is either very brave or very foolish. She is both intuitive and practical: intuitive in sensing that the WHITE WITCH will try to use EDMUND as bait to catch the other children; practical in taking pains to determine just how much Edmund knows to tell the witch and how long it will take her to catch them. Beloved by the children, she receives many gifts and honors at their crowning. Finally, along with Mr. Beaver, she is part of the happy company at the GREAT REUNION (LB 16).

BELIEF(S), BELIEVE — See FAITH; PROFESSION(S) OF FAITH.

BERN — A fine-looking man with a beard, this lord—one of the SEVEN NOBLE LORDS—is first seen on the island of Felimath, where he buys CASPIAN's freedom from the slave trader Pug. When he learns Caspian's identity, he immediately acknowledges him as KING and explains that he disapproves of the piracy and SLAVE TRADE on the island and has petitioned GUMPAS that they be stopped. An ideal nobleman, he has a gracious wife and merry daughters; and his people are free. He tells Caspian that he came to the islands weary of travel, fell in love, and settled down. His favors to the king earn him the gratitude of the Crown and the dukedom of the Lone Islands. He takes the OATH of office by placing his hands between Caspian's—a gesture of fealty from the time of chivalry and indicative of the noble and courteous class that Bern represents (VDT 2 and 7).

BIBLICAL ALLUSION(S) — Lewis had a lifelong appreciation of the Bible. In grammar SCHOOL, he read the Bible as a devout Christian; only when he had become an atheist in his early teens did he leave off his Bible reading. As a student of classical and English literature, he became a student of biblical content and style but rejected its claim to be inspired by God. When he recovered his Christian FAITH in his early thirties, he resumed his PRAYERful reading of the Bible and prayed the Psalms daily for the rest of his life.

There is only one explicit reference in the *Chronicles of Narnia* to the Bible as such: Reading it was not encouraged at EXPERIMENT HOUSE; therefore, EUSTACE doesn't even know of OATHS sworn on the Bible (SC 1).

The *Chronicles* are filled with biblical allusions—not direct or explicit scriptural references, but indirect hints of actual biblical phrases or suggestions of biblical themes or scenes. The following list is arranged in the order in which such allusions appear in each of the chronicles in their order by essential completion. The suggested allusion or reference is given, abbreviated, followed by the chapter number, and then the biblical parallel is cited.

The Lion, the Witch and the Wardrobe

Daughter of Eve (2)	Romans 5:12
I should live to see this day (7)	Luke 2:30
Wrong will be right when (8)	Matthew 12:18–20
At the sound of his roar (8)	Hosea 11:10–11

Sorrows will be no more (8)	Isaiah 65:19
When Adam's flesh and Adam's bone (8)	Genesis 2:23
They are tools, not toys (10)	Ephesians 6:11–17
No need to talk about what is past (13)	Isaiah 65:16
He just went on looking at Aslan (13)	Hebrews 12:2
Deep Magic (13)	1 Corinthians 2:7–8
I should be glad of company tonight (15)	Matthew 26:38
I am sad and lonely (15)	Matthew 26:38
Let him first be shaved (15)	Matthew 27:28
Jeering at him saying (15)	Matthew 27:29
In that knowledge, despair and die (15)	Matthew 27:46
Warmth of his breath . . . came all over her (16)	John 20:22
A magic deeper still (16)	1 Corinthians 2:7–8
ASLAN provided food (17)	John 6:1–14
He has other countries to attend to (17)	John 10:16

Prince Caspian

The People That Lived in Hiding (5)	Isaiah 9:1
Help may be even now at the door (12)	Mark 13:29
A few join his company (14)	John 6:66
Not water but the richest wine (14)	John 2:9

The Voyage of the *Dawn Treader*

As bad as I was (7)	James 5:16
Well—he knows me (7)	1 Corinthians 13:12
Caspian obeyed (13)	Ephesians 5:21
A little live coal (14)	Isaiah 6:6
Come and have breakfast (17)	John 21:12

The Horse and His Boy

Not the breath of a ghost (11)	Luke 24:39
Tell me your sorrows (11)	1 Peter 5:7
Joy shall be yours (14)	Matthew 25:21
Touch me (14)	John 20:27–29
Aslan was among them (15)	John 20:19
Not a Donkey! (15)	Daniel 4:24–33

The Silver Chair

If you are thirsty . . . (2)	John 4:10, 13–15
I have swallowed up . . . (2)	Psalm 21:9
There is no other stream (2)	John 7:37–38
Do so no more (2)	John 8:11
Remember the signs (2)	Deuteronomy 6:4–9
Aslan will be our good Lord (13)	Romans 14:8
Commend yourself to the Lion (13)	Psalm 31:5
I will not always be scolding (16)	Psalm 103:9
A great drop of blood (16)	1 John 1:7
It turned into a fine new riding crop (16)	Exodus 4:4
His golden back (16)	Exodus 33:23

The Magician's Nephew

Stars themselves . . . singing (8)	Job 38:7
It laughed for joy (8)	Psalm 19:5
Land bubbling like water (9)	Genesis 1:24
For out of them you were taken (10)	Genesis 3:19
Adam's race has done the harm (11)	1 Corinthians 15:21
Name all these creatures (11)	Genesis 2:19
My son, my son (12)	2 Samuel 18:33
Well done (14)	Matthew 25:21
Face to face . . . absolutely content (14)	Hebrews 12:2
Oh, Adam's sons . . . good (14)	Luke 19:42

The Last Battle

Trembled with a small earthquake (1)	Mark 13:8
Worst thing in the world (2)	Psalm 77:10
Is not like the Aslan (3)	Psalm 77:10
By whose blood (3)	Ephesians 1:7
Seeing is believing (10)	John 20:25–29
Between the paws of the true Aslan (10)	Deuteronomy 33:27
Lovely fruit trees (12–13)	Revelation 22:2
A Stable once had something inside (13)	Luke 2:7
Well done (13)	Matthew 25:21
The downpour of stars (14)	Mark 13:25

Aslan had called them home (14)	Baruch 3:34
Moon . . . looked red (14)	Joel 2:31
To know more of him (15)	Philippians 3:10
To look upon his face (15)	Psalm 27:8
Though he should slay me (15)	Job 13:15
No one got hot or tired (15)	Isaiah 40:31
One can't feel afraid (16)	1 John 4:18

BOGGLES — Evil spirits summoned by the White Witch to the Stone Table for the slaying of Aslan (LWW 13). Boggles are phantoms causing fright, goblins, bogeys, specters of the night, or undefined creatures of superstitious dread.

BREE — The strong, dappled stallion who carries Shasta to Archenland in HHB, his full name is Breehy-hinny-brinny-hoohy-hah, which sounds like a horse's neigh. A talking Narnian horse, he was (like Hwin) captured at an early age and taken to Calormen. To survive, he has had to hide his true nature. He longs for "Narnia and the North" and despises the Calormenes. It is from Bree that we learn the most about Narnian Talking Beasts, and he is an interesting combination of horsey attributes and human weaknesses.

Bree sees everything from the perspective of a horse: He remarks about the absurdity of human legs; he only begins to understand Shasta when he can perceive him as a human "foal" in need of riding lessons; and his sharp horse's hearing tells him immediately that the sound of hooves in the distance belongs to a thoroughbred mare

While Nikabrik and Trumpkin argue about his fate, Caspian awakes to the smells of a sweet, hot drink being offered him by Trufflehunter, the Talking Badger. (PC 5)

being ridden by a competent rider. His greatest weaknesses, his pride and his VANITY, are human.

BULLYING — See GANG, THE; SCHOOL(S).

BULVERISM — *Bulverism* is Lewis's coined term to describe the process of suggesting that another person's reasoning cannot be trusted or respected by calling attention to that person's motivations (e.g., "You say that because you're an atheist" or "You say that because you're a Christian") rather than arguing the merits of the issue.[1]

[1]For a further explanation see his very humorous essay "'Bulverism': or, The Foundation of 20th Century Thought" in *God in the Dock,* 271–277. See also REDUCTIONISM.

CAIR PARAVEL — The castle of the KINGS and queens in the GOLDEN AGE OF NARNIA, and the city in which it is located.[1] The castle sits high on a hill, overlooking the GREAT RIVER valley, on a seacoast peninsula near the mouth of the river. The chief mole, LILYGLOVES, planted an apple orchard for King PETER outside the north gate. In its prime, the great hall has an ivory roof, colored pavement floor, and tapestry-covered walls. The west DOOR is hung with peacock feathers and the east door is open to the sea (and, symbolically, to ASLAN, who comes from over the sea). It is not clear who, if anyone, lived in the castle before the Pevensie children; Mr. BEAVER recites an old rhyme that prophesies that the EVIL* TIME will end when Adam's flesh and bone sits on the throne at Cair Paravel. There are, in fact, four thrones already waiting at the castle, and it is there that Aslan crowns the kings and queens.

By the time of Prince CASPIAN, however, the castle and its surrounding city are greatly changed. TELMARINES are apparently responsible for having dug a channel across the peninsula, and the hill on which the castle sits is now a densely forested island. These are the GREAT WOODS (the Black Woods to the Telmarines, who think the woods are haunted by GHOSTS). The apple orchard has grown wild—right up to the north gate—but still bears delicious fruit. The castle itself is in ruins, and ivy has overgrown the door at one end of the dais that leads to the treasure chamber.

[1]The name "Cair Paravel" is probably from *kaer,* which is an old British word for "city" and *paravail,* from the Old French *par aval,* meaning "down," and Latin *ad vallem,* "to the valley." Thus, Cair Paravel is a "city in the valley" and takes its name from its castle.

CALDRON POOL — The source of the GREAT RIVER in Western NARNIA and the home of MOONWOOD the Hare, PUZZLE, and SHIFT. It is created by the GREAT WATERFALL, which churns like water boiling in a pot—hence its name (LB I and 8).

CALORMEN, CALORMENE(S)[1] — The empire to the south of ARCHENLAND and NARNIA, from which it is separated by the great desert. TASHBAAN is the capital city and the seat of power of the TISROC and his prime minister, the grand vizier. The lords and ladies of the realm are called TARKAANS and Tarkheenas and live in perfumed splendor. Lewis probably fashioned the name "Calormen" from the Latin *calor,* meaning "heat, warmth," and the English *men;* thus Calormenes are "men from a warm land." Indeed, the culture is modeled on the desert cultures of the Near and Middle East.[2] Lewis describes the Calormenes as a "wise, wealthy, courteous, cruel, and ancient" people. They worship the bloodthirsty GOD* TASH and hold life to be cheap: Slavery is common, war is a way of life, and citizens as well as slaves may be put to DEATH for minor offenses. Truly imperial, the Calormenes are determined that the free Narnians and Archenlanders shall someday submit to their rule, and from the GOLDEN AGE to the LAST BATTLE they do not give up their fanatical obsession with tyranny. The only named Calormenes to enter ASLAN'S COUNTRY are EMETH (who loved Tash with the love that was really ASLAN's) and ARAVIS (who sought freedom in the north). From Aslan's country LUCY can see the real Tashbaan, which suggests that something in Calormene culture is good enough to allow it to survive into eternity. Most Calormenes, however, are the archenemies of the free Narnians, and Lewis has given them qualities diametrically opposed to those of the Narnians.

Animals — There are no native TALKING BEASTS in Calormen. When such ANIMALS are kidnapped (notably the talking HORSES BREE and HWIN), they are treated badly. Ownership of animals is common practice in Calormen, and the Calormenes think nothing of over-riding their mounts, whipping and harnessing horses for hard work, and selling worn-out horses to be rendered into dog-meat. In Narnia, respect for life extends even to dumb animals. Talking horses are never mounted, except in battle, and are certainly not used as slaves.

Art and Poetry — The statues of gods and heroes that line the finer streets of Tashbaan are "impressive rather than agreeable to look at." In Narnia, craftsmanship is valued, and although its art is not specifically described, carved woodwork and medieval-style tapestries are mentioned. When Aravis and SHASTA hear that poetry will be sung at the Archenland feast, they prepare themselves to be bored, thinking northern verse consists of dull proverb after dull proverb. But when they hear the first bit of music, "a rocket seemed to go up inside their heads." When the song is over, they want to have this fireworks experience again.

Cities — Tashbaan is crowded with slaves, disorder, and filth. By comparison, the streets of Cair Paravel are clean, bright, and full of happy people. Green lawns gleam in the sunlight.

Class System — The line between the classes in Calormen is clear and abrupt, and the authority structure is rigid. The Tarkaans and Tarkheenas live in extreme COMFORT, take perfumed baths, and vacation at seaside resorts. They dress fabulously and lord it over the peasants, who live in filthy cities, dress poorly, and are generally dull and practical. They are impersonal and even cruel to their servants, to whose fates they are indifferent.

That there are classes in Narnian society is indisputable (given the monarchy and the existence of lords and ladies), but the differences are not highlighted. The peasant class of Narnia is made up of simple, hard-working folk—animals, DWARFS, and human beings alike. The assumed dignity of all Narnians eliminates the possibility of the condescending behavior toward assumed inferiors that permeates Calormene society. The class system can also be seen in formalized words of address: Calormene soldiers call their officers "My Master," and junior officers call senior officers "My Father." The Calormene tendency toward superlatives ("Tash the inexorable, the irresistible," "The Tisroc, may he live forever") also indicates the clearly drawn line between classes. There are no such Narnian forms of address except the polite "sir," which is part of Narnian COURTESY.

Clothing — Calormene clothing is based on Middle Eastern–style dress. The wealthy Tarkaans and Tarkheenas wear turbans and ROBES that are embroidered, bejeweled, and dripping with tassels and ornaments. Calormene peasants wear dirty clothes, turbans, and shoes that curl up at the toe. Narnian clothing is simple; it not only looks good but also feels good and SMELLS good. It is designed for freedom of movement and colored in earth tones.[3] Calormene battle-dress is also more opulent: Tarkaans wear chain-mail shirts (which TIRIAN calls "outlandish gear") and spiked helmets that poke through their turbans and carry curving scimitars, embossed shields, and lances. The Narnian chain-mail apparently weighs less, and Narnian soldiers carry straight swords.

Food — The food of Calormen is the food of desert feasts, and the Calormenes eat well and heartily, to the point of overeating. They cook with oil, onions, and garlic, while the Narnians use butter. Shasta's two meals in HHB succinctly point out the differences between Calormene and Narnian meals. His Calormen-style meal consists of lobsters, salad, snipe stuffed with almonds and truffles, rice with nuts and raisins, melons, desserts, and ices, all washed down with WINE. His Narnian meal in the Dwarfs' house is typical English fare: eggs, bacon, toast, milk, and coffee. Although Lewis is clearly partial to the Narnian food, the hungry Shasta enjoys both meals.

Hospitality — In Calormen, HOSPITALITY is demanded; in Narnia, it is willingly extended.

Physical Appearance — The Calormenes are a southern, desert race, dark and swarthy; they wear long beards, which may be perfumed, oiled, and dyed. The Narnians are very definitely northerners. They are mostly fair, pale-skinned people, and beards are worn mostly by older men and Dwarfs.

Religion — The Calormene religion is polytheistic. Three gods are mentioned (although more can be assumed): Tash, Azaroth, and ZARDEENAH. The Narnians are monotheistic and follow Aslan. Although nature gods abound in Narnia, they are alive, equal to the other Narnians, and not worshiped.

Slavery — The SLAVE TRADE is common in Calormen and is probably one reason for the wealth of the Tarkaans. Slaves are treated as badly as the animals and may be put to death for any assumed offense. In LB, the atrocity of Calormene slavery is in full force, and even Dwarfs and Talking Beasts are intended for labor in the Tisroc's mines. Slavery is abhorrent to the Narnians. When CASPIAN encounters the Lone Islands' thriving slave trade (which was allowed to flourish only because the islands were isolated from Narnian influence for so many years), he abolishes it immediately.

Storytelling — The Calormenes are consummate storytellers. Young Calormene schoolchildren learn storytelling as we learn essay-writing, and their stories are much more interesting than our essays. Both Aravis and Emeth tell their STORIES in the grand Calormene fashion. Narnians also love stories, but apparently—like the Narnians themselves—these stories tend to be more truthful and less filled with embellishment.

Women — Calormene women are chattels of their husbands and parents. Their marriages are arranged and their own FEELINGS are irrelevant. They do not know how to write. Narnian women seem to be equal to men, and to enjoy the same rights and privileges.[4]

[CURRENCY.]

[1]Accent on the middle syllable ("ca-LOR-men," "ca-LOR-mean") according to Douglas Gresham, Lewis's stepson.

[2]See RACISM AND ETHNOCENTRISM.

[3]See ROBE(S), ROYAL.

[4]See SEXISM.

CAMILLO — A talking hare who leads a contingent to the great council (PC 6 and 7). Just before the council begins, he warns of a man nearby, who turns out to be Doctor CORNELIUS.

CARTER — One of the students in power (likely a member of the GANG) at EXPERIMENT HOUSE, he seems to have delighted in causing ANIMALS to suffer PAIN. Eustace confronts him in a matter concerning a rabbit (SC 1).

CASPIAN I — The first TELMARINE* KING of NARNIA (PC 4), also known as "the Conqueror." He brings the Telmarine nation into Narnia in 1998 N.Y.[1] and fights against the NINE CLASSES OF NARNIAN CREATURES. His armies silence them, kill them, drive them away, and try to erase their MEMORY from the land because they are the subjects of ASLAN, whom the Telmarines FEAR.

[1]*The Land of Narnia*, 31.

CASPIAN VIII — A TELMARINE* KING of NARNIA, father of CASPIAN IX and MIRAZ (PC 13).

CASPIAN IX — A TELMARINE* KING of NARNIA, son of CASPIAN VIII, brother of MIRAZ, and father of CASPIAN X (PC 5 and 13). He is murdered by Miraz in 2290 N.Y.[1]

[1]*The Land of Narnia*, 31.

CASPIAN X — A TELMARINE* KING of NARNIA (born 2290 N.Y., reigned 2303–2356 N.Y., died 2356 N.Y.) called "the Seafarer." He is the orphaned son of CASPIAN IX, raised by the usurper MIRAZ and his wife, PRUNAPRISMIA, husband of the DAUGHTER OF RAMANDU, father of RILIAN the Disenchanted. [SPOILERS] Caspian is the hero of the great WAR OF DELIVERANCE in which Old Narnia is freed from the tyranny of its Telmarine conquerors, who style themselves New Narnians. Caspian is also leader of the voyage of the *DAWN TREADER,* a QUEST that he undertakes with ASLAN's permission in order to discover the fates of the SEVEN NOBLE LORDS who stood faithfully by his assassinated father, as well as to explore the islands of the Eastern Sea. Caspian's years as king are marred by the horrible DEATH of his beloved wife and the disappearance of his only son and heir. But these hurts are mended by the mercy of Aslan. His story is one of the most complete that Lewis tells in the *Chronicles,* and his is the best-developed character of all the Narnians.

CAT(S) — The big cats in the *Chronicles* exhibit the full range of feline characteristics: cunning and ferocity combine with independence and VANITY. Because Lewis loved cats and "owned" several, they almost always use these characteristics in ASLAN'S NAME.[1]

The first cats to be created—"panthers, leopards and things of that sort"—are eager to use their new muscles in chasing down ANDREW. When Andrew is cornered, he is terrified to see "cool-looking leopards and panthers with sarcastic faces." In LWW, when, under the fluttering BANNER of a red rampant lion, ASLAN appears in glory, one leopard bears his crown and another his standard (LWW 12). Swift and good strategists, great cats are fierce in battle (LWW 13). For their part in the WAR OF DELIVERANCE, leopards are honored and rewarded at the coronation of KINGS and queens. In HHB, at the battle at ANVARD (13), the great cats form a circle and attack the CALORMENE* HORSES, then defeat the rams-men to greatly advance the Narnian forces. When TIRIAN finds that the TREES in LANTERN WASTE are being felled, he sends for some skilled warriors, including "a leopard or so" (LB 2). Cats can also be great sources of reassurance, warmth, and protection, as we see in the ancient tombs when SHASTA discovers Aslan in the form of a cat (HHB 6).

[1]"We were talking about cats and dogs the other day and decided that both have consciences but the dog, being an honest and humble person, always has a bad one, but the cat is a Pharisee and always has a good one. When he [the cat] sits and stares you out of countenance he is thanking God that he is not as these dogs, or these humans, or even as these other cats!" (*Letters to an American Lady,* 21 March 1955. See also 24 February 1961.) See *Past Watchful Dragons,* 16.

CENTAURS — In Greek and Roman MYTHOLOGY, a semidivine being with the head and chest of a man and the body of a HORSE. For Lewis the Centaur represents the harmony of nature and spirit.[1] In the *Chronicles,* the Centaur is one of the NINE CLASSES OF NARNIAN CREATURES. Four great Centaurs are described in LWW: Their horse parts are "like huge English farm horses"; the human parts are "like stern but beautiful GIANTS" (12). In PC 6, we are told that a Centaur's diet consists of oaten cakes, apples, herbs, WINE, and cheese. They use both swords and hooves in battle and are good strategists in warfare. They are to be respected: According to TRUFFLEHUNTER, no one ever laughs at a Centaur (PC 13), and in SC 16 the CHILDREN ride Centaurs bareback, because "no one who valued his life would suggest putting a saddle" on one. They are informed about the properties of herbs and roots and astrology. In HHB 14, a Centaur has predicted that Cor will save ARCHENLAND from its greatest danger. The three named Centaurs in the *Chronicles* are GLENSTORM, CLOUDBIRTH, and ROONWIT; their names have to do with weather, wildness, and wisdom. Other Centaurs in the *Chronicles* include two stone Centaurs in the courtyard of the castle of the WHITE WITCH. Centaurs are in the war party that marches against RABADASH in HHB 12 and in the crowd at Cair Paravel in SC 3.

[1]*Miracles,* seventh ¶ from the end of Chapter 14 and fifth ¶ from the end of Chapter 16. See also *The Great Divorce.*

CHARN — The dead city, and likely the world, of which Jadis is empress. Even its SUN is dying, and a dull reddish light hangs over the blackish blue sky. It is a cold, SILENT world where nothing lives. The very name of the city has a sinister sound and suggests burning. Lewis may have appropriated the name "Charn" from *charnel,* meaning "a burial place."[1] [MN SPOILER] Jadis destroyed all living things in Charn with the DEPLORABLE WORD in an attempt to usurp her sister's power. The frozen figures of Charn's royalty that DIGORY and POLLY encounter in the great hall show the degeneration of the KINGS and queens from dynasty to dynasty, culminating in Jadis herself. The last years of Charn were ones of great cruelty—slavery, human sacrifice, and endless warfare. The pool of Charn in the WOOD BETWEEN THE WORLDS has dried up; according to Aslan, "That world is ended, as if it had never begun" (MN 15). He adds that SONS OF ADAM and Daughters of Eve should take that as a warning, and it is clearly intended as a warning to the reader that our world, too, may be destroyed through unchecked despotism and the use of weapons of mass destruction.

[1]Professor T. W. Craik, letter to the author, July 31, 1979. For more, see WORLD ASH TREE.

CHERVY THE STAG — Chervy is a beautiful ANIMAL, regal and delicate in appearance (HHB 12). Unlike the smaller animals, this TALKING BEAST is not indecisive when he hears from SHASTA of the approach of RABADASH and his cohort to ARCHENLAND and NARNIA. He bounds away and meets EDMUND, SUSAN, and CORIN, newly arrived from their embassy to TASHBAAN. *Chervus* is the Latin word for "deer."

CHIEF VOICE — Also known as the Chief Duffer and the Chief Monopod, he is the leader of the DUFFERS, their spokesman, and, indeed, does all their thinking—such as it is (VDT 9).

CHILD, CHILDREN — Children and YOUTHS are the heroes and heroines of the *Chronicles* who are brought to Narnia time and again to clean up the messes made by ADULTS. Lewis takes them very seriously—more seriously, in fact, than he takes the grown-ups. Upon entering NARNIA, children become subtly older and more mature, but they retain their childlike innocence, candor, and knowledge of RIGHT AND WRONG. As characters, they are far more realistically drawn than any grown-up in or out of Narnia. Lewis had not forgotten what it was like to be a child, and the STORIES are told from a child's point of view. Children are often much more intelligent than grown-ups think they are, and they can use this to their advantage. Lewis takes children's night FEARS very seriously and involves himself and his readers in SHASTA's experience of the howling jackals at the tombs. The first chapter of MN is full of things that children like to do: build forts, look for treasure, and explore hidden places—especially haunted ones.

Lewis was aware that many children are left to their own devices at an early age, an AUTOBIOGRAPHICAL ALLUSION. [SPOILERS] So Shasta is motherless and lives with ARSHEESH, who is not his real father; Prince CORIN has a father, King LUNE, but he has been motherless from an early age; ARAVIS has a mean stepmother and no affection for her father, who wants to marry her off to the tiresome AHOSHTA; CASPIAN lives with his Uncle MIRAZ, who intends to kill the prince after his own son is born; DIGORY's father is in India, having left him with his seriously ill mother and his insane Uncle ANDREW; RILIAN's mother is the idealized DAUGHTER OF RAMANDU, dead from the bite of a serpent; EUSTACE's parents, ALBERTA and HAROLD, are totally lacking in IMAGINATION; and the Pevensie children love their parents, but they are left on their own a lot of the time.

Lewis tries to communicate through the *Chronicles* the importance of remaining child*like*—as opposed to child*ish*—in outlook. [SPOILER] SUSAN, the only "failed child" in Narnia (she might almost be called a "lapsed child" in the way the LAPSED BEAR OF STORMNESS is a failed bear) has fulfilled Lewis's dictum that in our world, too, "it is the stupidest children who are the most childish and the stupidest grown-ups who are the most grown-up." Hers is a false maturity: "She wasted all her SCHOOL time wanting to be the age she is now, and she'll waste all the rest of her life trying to stay that age."

CHOLMONDELEY MAJOR — See GANG, THE.

CHORIAMBUSES — See CORIAKIN.

CHRONOSCOPES — See CORIAKIN.

CLIPSIE — Daughter of the CHIEF VOICE, she recites the MAGIC spell of invisibility in the magician's book, which makes the DUFFERS disappear. Her father estimates that she is the same age as LUCY (VDT 9).

CLODSLEY SHOVEL — A mole to whom Lewis humorously assigns the name of British naval hero Admiral Sir Cloudesley Shovel. This TALKING BEAST leads a troop of moles to the great council, where he proposes throwing up entrenchments around DANCING LAWN as a first line of defense (PC 6). After the WAR OF DELIVERANCE, he directs his moles to prepare the various earths for the feast of the TREES (PC 15).

CLOUDBIRTH — A CENTAUR and famous healer among the Narnians in the later reign of CASPIAN X. As a

Trufflehunter watches as the Three Bulgy Bears give Caspian the kisses he deserves as a Son of Adam and future king of Narnia. (PC 6)

Centaur he has knowledge of the healing properties of herbs and plants, and he is sent to minister to PUDDLEGLUM's burnt foot (SC 16).

COL — According to Lewis MS 51,[1] in 180 N.Y. Prince Col leads a group of Narnians to settle the uninhabited country of ARCHENLAND. He is the younger son of King Frank V of Narnia and becomes the first KING of Archenland. In MN, however, Lewis says that the first king of Archenland is the second son of King FRANK and Queen HELEN (15).

[1]*The Land of Narnia*, 31.

COLNEY 'ATCH — A voice in the crowd at the LAMP-POST calls Jadis the "Hempress of Colney 'Atch." Colney Hatch was a London insane asylum at the turn of the last century. Thus in 1900, "Colney 'Atch for you!" was Cockney slang for "You're crazy!" (MN 8).

COMFORT — Comfort and safety, Lewis seems to say, are very great dangers in life because they can instill a false sense of security. In SC, it is comfort at HARFANG and irritation with lack of comfort on ETTINSMOOR that cause JILL to forget the SIGNS. It is in HHB 12, set in the GOLDEN AGE, that questions of comfort first arise. The smaller woodland ANIMALS are so secure that "they are getting a little careless." LASARALEEN Tarkheena is a slave to creature comforts; and although Aravis is somewhat tempted by luxury, she is glad to escape to the more austere north. SHASTA is so comfortable in the Narnian embassy that "none of [his] worries seem so pressing" (4 and 5). This gives him a good feeling but a false sense of security. His dinner and conversation with TUMNUS are so enjoyable that they cause him to squelch his worries about ARAVIS and BREE and hope that he can stay long enough to be taken to NARNIA by ship instead of having to brave the terrors of the great desert.

CONSCIENCE — See RIGHT AND WRONG.

COR — See SHASTA.

CORDIAL — FATHER CHRISTMAS'S MAGIC gift to LUCY in LWW (10). A strong, sweet liquor (see WINE), it is made from the juice of the FIRE-FLOWERS that grow in the valleys of the SUN, is kept in a diamond bottle, and is said to be able to heal almost every wound and illness.

CORIAKIN — A MAGICIAN and master of the land of the DUFFERS, he is an old man with a waist-length beard, barefoot, dressed in a red ROBE and crowned with a circlet of oak leaves. His staff is strangely carved. According to RAMANDU, Coriakin was once a STAR, and his governing of the Duffers is a kind of punishment for having failed in some way. He longs for the day when he can rule the Duffers by wisdom rather than by rough ways. He has a room full of polished instruments: astrolabes (instruments to determine altitudes and for solving other problems of practical astronomy); orreries (mechanisms devised to represent the motions of the planets about the SUN by means of clockwork); chronoscopes (instruments for observing and measuring time); poesimeters (imaginary instruments for measuring the meter of poems); choriambuses (imaginary devices for measuring choriambs—a metrical foot consisting of four syllables: long, short, short, long); and theodolinds (portable surveying instruments used in measuring horizontal angles—theodolites) (VDT 9–11 and 14).

CORIN THUNDER-FIST—[HHB SPOILERS] SHASTA's twin brother, Prince Corin of ARCHENLAND (HHB 4, 5, 12, and 13). He is twenty minutes younger than Cor and hence second in line for the throne of his father, King LUNE. RILIAN sings an old song about his sometimes VIOLENT exploits (SC 13). SUSAN has been his best friend since his mother died, and he and Shasta look so alike that Susan mistakes Shasta for Corin. He is a feisty boy, fond of fights, and to defend Susan's HONOR he knocks down a boy in TASHBAAN. Full of a sense of honor, he is offended at Shasta's suggestion that he would tell King EDMUND and Queen Susan anything but the truth. That he and Shasta become instant friends is indicative of their then-unknown but strong blood bond. He enters the battle of ANVARD against his father's wishes—fighting Thornbut to do so—and forces Shasta to suit up in the DWARF's armor and join him in battle. Corin is scolded for his rashness by King Lune, but the KING can't disguise his pleasure at his son's COURAGE. He is the most incensed of the lords at RABADASH's insult of the king and taunts the CALORMENE, for which act he is rebuked by Lune. He rejoices that his brother will be king instead of him, because he knows he will have fun as a prince while Cor must shoulder the responsibility of his VOCATION. Corin grows up to be the best boxer in Archenland (HHB 15) and earns the surname "Thunder-Fist" in a thirty-three-round fight in which he wins the LAPSED BEAR OF STORMNESS back to the ways of TALKING BEASTS, an UNFINISHED TALE of NARNIA.

CORNELIUS, DOCTOR — CASPIAN X's beloved tutor (PC 4), a half-DWARF who keeps the facts of his heritage hidden during the FEAR-ridden rule of MIRAZ. It is easy to keep his identity hidden from the New Narnians, who know nothing of what Dwarfs look like. To them he is merely a very small, fat SCHOOLteacher with a long, pointed silver beard. His wrinkly brown face looks at once wise and ugly and kind. It is Doctor Cornelius who enlightens the young prince as to the real HISTORY of NARNIA. He longs for the return of Old Narnian values, even though he fears that his Dwarf relatives might despise him because of his human blood. Among the subjects Doctor Cornelius teaches are history, grammar (for which he uses a book written by Pulverulentus Siccus), and ASTRONOMY. Because he teaches the truth about the world, he is obviously the sort of educator Lewis respects, and is certainly so in contrast to the HEAD of EXPERIMENT HOUSE. His revelation to Caspian that he is a half-Dwarf is a NUMINOUS experience for the boy. When Caspian is king, he makes Doctor Cornelius his lord chancellor.

COURAGE — True courage[1] is the strength of ASLAN, and REEPICHEEP is its embodiment. Because he has neither HOPES nor FEARS, his courage in the face of danger is never in question. For human beings, however, courage is a more complex question and their hopes and fears often cause them to waver (RILIAN tells EUSTACE, JILL, and PUDDLEGLUM to bid farewell to hopes and fears [SC 13]). As

Caspian says to Reepicheep, "There are some things no man can face" (VDT 13). In PC 10, Lucy tries to avoid Aslan's look by hiding her face in his mane, but the magic in his mane fills her with lion strength and he calls her a lioness. Miraz has false courage, which leads him to accept the monomachy with Peter that ends in his death (PC 13). In VDT 8, Eustace acts bravely "for the first time" in his life when he attacks the Sea Serpent. And his volunteering to stay overnight on Ramandu's Island (VDT 12) is especially brave. In HHB 13, Prince Corin's courage is not in question but his obedience is. However, King Lune and Darrin seem to speak for Lewis in praise of a rash courage over a planned cowardice. Shasta's courage in the face of the lion shames Bree (HHB 10).

[1]See *Mere Christianity,* Book III, Chapter 2.

COURTESY — Narnians show great respect and hospitality to one another. In LWW 12, Mr. Beaver defers to Peter (Sons of Adam before animals[1]); Peter says to Susan, "Ladies first"[2] (albeit out of fear), and Susan replies, also out of fear, that he should go first because he is the eldest. In PC, we learn that it is bad manners among squirrels to watch anyone going to his store or to look as if you wanted to know where it is. Peter defers to Trumpkin as the eldest in the vote to go upstream or downstream. Nikabrik shows his true colors when he relegates his oath of allegiance to Caspian to "court manners," which he feels can be dispensed with considering the tight situation they are in. In VDT 2, Reepicheep teaches Eustace some manners, respect for knighthood, and respect for mice and their tails. He also reprimands Rhince for saying "good riddance" to Eustace (VDT 6), because (1) Eustace is of the same blood as Queen LUCY, and (2) he is a member of their fellowship and thus it is a matter of HONOR to find him or avenge him. The entire company stand and uncover their heads in the presence of RAMANDU and his DAUGHTER because they are "obviously great people" (VDT 13). The life of the SEA PEOPLE includes courtesy along with peace, rest, and council (VDT 14). Out of courtesy, DRINIAN offers Caspian a drink of the sweet waters before he takes his own taste (VDT 15). In SC 3, the reformed Eustace is impressed with the glory and courtesy of the Narnian supper. GLIMFEATHER courteously offers to catch a bat for Jill as a snack (SC 4). When UNDERLAND is free, all bow to RILIAN (SC 15), and he in turn is deferential to the eldest DWARF, whom he calls "Father." SHASTA (raised in CALORMEN, where genuine courtesy is not practiced) was never taught not to listen behind DOORS (HHB 1). Because RABADASH has not exercised the courtesy of nations (in which he ought to have declared war on ARCHENLAND by sending defiance), King LUNE declares him unworthy to fight a man of honor (HHB 15). In LB 11, TIRIAN* SILENCES Eustace's scolding of the Dwarfs: "No warrior scolds. Courteous words or else hard knocks are his only language."

[1]See HIERARCHY.

[2]See SEXISM.

COWARDICE — A quality despised by all true Narnians, who value COURAGE. Cowardice is distinguished from FEAR, which is a FEELING, and therefore natural and acceptable. Rather, it is the yielding to the temptation of fear that leads to cowardice, and thus away from the ADVENTURE that ASLAN sends. The DUFFERS' fear of the dark and the invisible MAGICIAN has led to a cowardice so great that they won't ask their own sons or daughters to break the spell (VDT 9). PITTENCREAM, who refuses to go with the rest of the DAWN TREADERS on their voyage to the Utter East, is the embodiment of cowardice (VDT 14). His refusal to meet the adventure leads to his being cast out from the group and to his degeneration as a teller of lies in CALORMEN. In MN 2, DIGORY accuses Uncle ANDREW of cowardice, and Andrew turns the accusation back on his nephew, expressing the HOPE that Digory won't show the white feather.[1]

[1]"To show the white feather" is a phrase from cockfighting (gambling on which rooster will kill the other). The white feather in a gamecock's tail is a sign of degenerate stock and thus of cowardice.

CREATION OF NARNIA — ASLAN creates NARNIA with his wild, glorious song (MN 8 and 9). The STARS are first to appear, simultaneously and in harmony with Aslan, the First Voice; the Voice rises and with it the SUN rises over Narnia; its light illuminates the mountains, through which a river runs; and green grass spreads from Aslan himself to cover all of Narnia with plant life and TREES. As Aslan's song becomes more tuneful, yet wilder, ANIMALS spring forth from the earth, which bubbles and boils with their activity; birds shower from the trees, and bees and butterflies are already busy with the flowers; the TALKING BEASTS are separated from the DUMB BEASTS; and finally, as Aslan commands Narnia to awake and be divine, all the good MYTHOLOGICAL "peoples" are called forth.

CREEK(S) — See Glasswater Creek, creek(s).

CRUELS — Evil beings, summoned to and present at the slaying of Aslan (LWW 13). Their name reveals their characteristic behavior.

CRUELTY TO ANIMALS — See Pain.

CRY(ING) — In the *Chronicles,* crying is both ignoble and to be avoided, and noble and not to be ashamed of; it is often a sign of despair or self-pity, and it is also a reaction to great beauty or tragedy. While Lewis excuses crying, he also says that it is not the way to get anything accomplished. In SC, Lewis comments, "Crying is all right in its way while it lasts. But you have to stop sooner or later and then you still have to decide what to do." In LWW 12, Lucy and Susan cry in despair over Aslan's death, and in LB 3, 11, and 12 Tirian and Jewel shed "bitter tears" at the death of their hope in Aslan, and Jill cries over the death of the horses. There is much more crying in self-pity, however: SC opens with Jill crying behind the gym (1), and she has a good cry after Eustace plunges over the cliff (2). She later breaks down and cries from weariness at the castle of Harfang (6), and she is so troubled by her dream that she cries into her pillow (8). It is a sign of her growth that she tries to restrain herself from crying during the killing of the Queen of Underland (12), and later her only crying is for the beauty of Aslan and the sadness of Caspian's funeral music (15). When she notices Eustace crying, she observes that he's crying, not freely like a child or ashamedly like a boy, but like an adult (1). Eustace, too, has come a long way from the tears he shed as a DRAGON in VDT 6. In HHB 11, SHASTA feels so sorry for himself on his journey through the fog in ARCHENLAND that he cries. But at the frightening discovery of a large, breathing presence beside him, he stops crying "now that he really [has] something to cry about." In MN 1, DIGORY is so miserable that he doesn't care who knows he has been "blubbing," and later he cries for his mother again—but this time Aslan cries with him (12 and 14)! In LB 10, Jill is a much stronger character, and she feels like crying for JOY at EMETH's zeal and beauty. Finally, Jill, Lucy, and Tirian cry in mourning the passing of NARNIA (14).

CURIOSITY — In Lewis's thinking, curiosity is more than an eagerness and an aptitude for knowledge. It is also a power of the human mind that is open to abuse. In the medieval tradition,[1] curiosity (Latin: *curiositas,* "koor-ree-OH-see-tahs") is a form of the vice of intemperance, and studiosity (Latin: *studiositas,* "stew-dee-OH-see-tahs") is a form of the virtue of temperance. The worst distortion of this exaggerated striving for knowledge for Lewis and the medievals is MAGIC. In VDT 10, LUCY's desire to know what her FRIENDS think of her leads to a disturbing experience with the magician's book. POLLY's curiosity overcomes her caution and she steps into Uncle ANDREW's study (MN 1). When Andrew reveals that his godmother's box contained something from another world, DIGORY becomes "interested in spite of himself" (2). In the great hall of CHARN, his wild curiosity to know what will happen if he rings the GOLDEN BELL almost ruins them all (4). Digory's natural curiosity (3) might have taken him the way of his uncle, the

EVIL magician, but it did not. Instead he developed his *studiositas* and became Professor Kirke, a man concerned with the correct use of knowledge. The HERMIT OF THE SOUTHERN MARCH has also learned to control his curiosity, and he says of the future, "if we ever need to know it, you may be sure we shall" (HHB 10). The calmness and serenity of his enclosure embody the peace that this attitude has brought him.

[POSITIVITY; TECHNOLOGY.]

[1]See Josef Pieper's *The Four Cardinal Virtues* (Notre Dame: University of Notre Dame Press, 1966), 198–202, for a succinct treatment of this matter. See also *The Abolition of Man,* Chapter Three.

CURRENCY — Narnian Lions and Trees, everyday coinage used in Beruna, are discovered on DEATHWATER ISLAND (VDT 8). The CALORMENE crescent, chief coin of the Lone Islands (VDT 3), is worth about "one-third of a [British] pound." Another coin, the minim, is one-fortieth of a crescent (VDT 4, HHB 5). ANDREW Ketterley's three silver half-crowns and sixpence become the SILVER TREE and his two gold half-sovereigns become the GOLDEN TREE.

D

DANCE — Dances and dancing are the chief means of celebration in NARNIA.[1] In LWW 16, all the creatures ASLAN revives dance around him, and at the great coronation feast at Cair Paravel there is REVELRY and dancing (LWW 17). PC (6, 9, 10, 11, 14, and 15) features a number of dances, all of which take place at the DANCING LAWN. In VDT 15, dancing is said to be the leisure-time activity of the SEA PEOPLE. The Great Snow Dance in SC 15 is an annual Narnian event held north of the GREAT RIVER on the first moonlit night when snow is on the ground.

[1]Though Lewis was something of a TRUFFLEHUNTER at dancing himself (PC 6), the dance was for him the happiest image for the most important things in life: the place of the human race in the universe, the relationship of humankind with the earth and its creatures, the relationship among human beings (especially between the sexes), the relationship of humankind to God, and the inner life of God himself.

DANCING LAWN — The traditional site of feasts and councils in Old NARNIA, located west of the River Rush and south of the GREAT RIVER and ASLAN'S HOW. Elms border the smooth circle of grass, and there is a well at the margin (PC 6 and 10). It is the site of all DANCES in the *Chronicles* except the Great Snow Dance, which takes place north of the Great River almost directly above Bism (SC 15).

DARK ISLAND — [VDT 12 SPOILER] A place of terror where all DREAMS come true, including nightmares. For Lewis, the Dark Island seems to be the place of CHILDREN'S FEAR of the dark.[1] It is located fourteen days of gentle wind southeast of the land of the DUFFERS. As first sighted by Edmund, it appears to be a dark mass or mountain, but on closer view, it is seen to be utter blackness, like the interior of a tunnel or the edge of a night without moon or STARS. The waters around the island are greasy and lifeless, and it is impossible to ascertain the speed or direction of the *DAWN TREADER* in this murky sea. Piercing cold afflicts all but the toiling oarsmen as the ship moves to rescue Lord RHOOP. Exhausted and terror-stricken, Rhoop refuses to talk about the "they" who held him captive on the island, and he asks CASPIAN to respect his PRIVACY by promising that he will not question him further about his experiences. So impressed is Caspian by Rhoop's suffering that he gladly makes this promise and immediately directs the crew to sail away from the island as fast as possible. At this point, the darkness is so palpable that they expect the ship to be coated with grime, and total silence engulfs them. Yet each one hears a different SOUND of terror: a gong, giant scissors cutting, creatures crawling up the side of the ship, "it" landing on the mast. Finally, Rhoop screams in anguish that they will never get out. But after LUCY calls on ASLAN for help, a tiny speck of light appears and then a broad beam illumines the ship like a searchlight. An ALBATROSS (Aslan himself or his messenger) appears to lead the ship into the full light of day.

[1]So important did Lewis consider night-fears that he extensively revised the ending of this twelfth chapter for the pre-1994 American editions of VDT. His aim was to correct any impression that the original British edition might have given that night-fears are unreal and ultimately laughable and that they can be obliterated altogether. Thus in the pre-1994 American editions, the Dark Island and its darkness do not vanish but the size diminishes gradually as the *Dawn Treader* sails away. See DREAMS for a delineation of the differences between editions. See also *Letters to Children*, 33–34.

DAUGHTER OF RAMANDU—[VDT and SC SPOILERS] A tall, beautiful young woman with long golden hair, who greets the travelers when they arrive at RAMANDU'S ISLAND. She is dressed in a long blue gown, cut so as to leave her arms bare, and she carries a tall candle in a silver candlestick. Its flame burns with steady intensity as if in a closed room. She is known only by title (Ramandu's daughter, the STAR's daughter, CASPIAN's queen, RILIAN's mother) and by no personal name throughout the *Chronicles*—an indication of the awe with which Lewis wished to surround her.[1] She meets a horrible DEATH when she is bitten by a serpent while resting after going Maying with her son, Rilian. Her body is carried back to the city, where she is mourned as a gracious and wise lady in whose veins flowed the blood of stars. She is named among the faithful at the GREAT REUNION.

[1]For another view, see SEXISM.

DAWN TREADER, DAWN TREADER — A Narnian ship built for King CASPIAN and commanded by Lord DRINIAN. Shaped like a DRAGON's head, it has green sides and a gilded prow and stern. A square sail of rich purple is rigged to its one mast. Caspian's cabin is small but beautifully decorated in the Chinese style, with birds, beasts, vines, and crimson dragons on painted panels. A flat, gold image of ASLAN fills the space above the door. There are two long hatches, one fore and one aft of the mast, with benches for rowing. A pit, reaching down to the keel, lines the center of the ship and holds provisions.[1] The crew includes about thirty swordsmen.

When her journey to rescue the SEVEN NOBLE LORDS is complete and the adventurers make ready to return from WORLD'S END, the ship flies all her flags and displays all her

It is the worst of nights: The Giant Wimbleweather weeps tears of failure; the bloodied bears, the wounded Centaur, and all the other creatures huddle in gloom under the dripping trees. (PC 7)

shields to HONOR* REEPICHEEP, EDMUND, LUCY, and EUSTACE. Each of them shares the title "Dawn Treader," as do all those who journeyed to the World's End with King Caspian. This honor becomes their most precious bequest to their heirs.

[1]Most editions of VDT print Pauline Baynes's fine diagram of the ship. Look for it.

DEATH — Death is called the "long journey" in a conversation between ASLAN and CASPIAN's old NURSE (PC 14). In SC 13, RILIAN interprets the marvelous transformation of his shield as a SIGN that "Aslan will be our good lord, whether he means us to live or die. And all's one, for that." Later in SC (16), the old King Caspian dies and is reborn as the young Caspian in ASLAN'S COUNTRY. He is truly alive in Aslan's country, although he speculates that he might be a GHOST if he turned up elsewhere. In HHB 6, SHASTA is sure that the Lion is going to kill him, and he wonders if anything happens to people after they die. When FRANK comes into Narnia he thinks he has died, and his reaction is remarkable for its piety: He sings a hymn (MN 8). In MN, Lewis stresses the point that death is not the greatest tragedy. DIGORY does not want to be immortal; he would rather die and go to heaven (13). Later Lewis comments that "there might be things more terrible even than losing someone you love by death" (14). The characters in LB spend a lot of time discussing death. PUZZLE says the dead lion must be given a decent burial (1); the second mouse says to TIRIAN that "It would be better if we'd died before all this began" (4); ROONWIT's last words are about noble death (8); EUSTACE and JILL discuss their possible deaths in the LAST BATTLE (9); and JEWEL, in a discussion with POGGIN, Tirian, and Jill, is confident that death is the way into Aslan's country (12). And, indeed, it is. The STABLE* DOOR becomes a metaphor for death: On this side of the door death is terrifying, black, unknown; but on the other side lies the glory of Aslan's country.

DEATHWATER ISLAND — An island about twenty acres in size, the seventh landing of the *DAWN TREADER* (VDT 9, 11, and 16). Dominating the island is a mountain, at the top of which is a little lake that is twelve or fifteen feet deep and almost completely surrounded by cliffs. CASPIAN, REEPICHEEP, EUSTACE, EDMUND, and LUCY discover the rusted armor of one of the SEVEN NOBLE LORDS on the lakeshore. Then they see a beautiful golden statue at the bottom of the lake. Edmund deduces that the "statue" is really the body of the lord turned to purest gold by the deadly MAGIC of the lake waters. Edmund and Caspian prepare to fight over their GREED for the gold, until a vision of ASLAN wipes their MEMORIES clean of everything but the awareness that the island is dangerous and that one of the lords (they later decide it has to have been RESTIMAR) died there. Reepicheep names the island Deathwater.

DEDICATIONS, DEDICATEE(S) — Lewis's particular affection for the children of some of his dear friends is the source of the dedications of the first six chronicles (LB has no dedication).

LWW is dedicated to LUCY Barfield (1935–2003), Lewis's goddaughter and the adopted daughter of Owen and Maud

Barfield, who was four when Lewis began to write the book and thirteen when he resumed and finished it.[1] Lucy loved music and ballet and eventually taught music. In 1966 she was diagnosed as having multiple sclerosis. She married Bevan Rake and lived happily, although she was often hospitalized. When her husband died in 1990, her health deteriorated; she lived in the hospital for the rest of her life. During that time, she told Walter Hooper how much the dedication meant to her: "What I could not do *for myself*, the dedication did for me. My godfather gave me a greater gift than he could have imagined." Hooper wrote:

> As every creature comfort was taken from her, and she had lost her sight, Lucy's faith in God grew and blessed not only her, but also those who knew her. Owen Barfield, touched by her humility, said many times, "I could go down on my knees before my daughter." During the last seven years of her life in the Royal Hospital for Neurodisability in London, her brother Jeffrey—to whom Lewis dedicated the Voyage of the "Dawn Treader"—read her the *Chronicles* of Narnia.[2]

Mary Clare Havard (1936–) is the daughter of Lewis's doctor (who was a member of the Inklings). Lewis asked her to read the typescript of LWW and he thanked her with the dedication of PC.[3]

Mr. and Mrs. Barfield fostered Geoffrey Corbett (1940–, also known as Jeffrey Barfield), whose birth mother could not provide for him. Lewis paid Geoffrey's school fees, and he became first a landscaper, then a welder, then a lay evangelist. He took care of his foster parents until their death. He took their name in 1962.[4] Lewis dedicated VDT to him.

David (1944–) and Douglas (1945–) Gresham[5] are the sons of William Lindsay Gresham and Joy Davidman Gresham. When she fled to England in 1952 to avoid her abusive husband, Joy took the boys; Lewis was first their benefactor and then, when William divorced her and Lewis married her, became their stepfather. That he dedicated HHB to them is a sign of his care for them.

Nicholas Hardie (1945–), the son of the Inkling Colin Hardie and his wife, Christian Viola Mary Lucas, is the dedicatee of SC.

MN is dedicated to the friends of Lewis's poet-correspondent,[6] Mary Willis Shelburne: the Kilmer family of Washington, D.C., Kenton (father, son of the American poets Joyce and Aline Murray Kilmer) and Frances (mother) and eight of their ten children: Hugh, Anne, Noelie, Nicholas, Martin, Rosamond, Matthew, and Miriam (Deborah and Jonathan were not yet born). At the encouragement of Mrs. Shelburne and the children's grandmother, the children began writing Lewis in 1954, and he and they corresponded for a time thereafter.[7] Nicholas Kilmer writes:

> My father, Kenton Kilmer, had been poetry editor of the *Washington Post,* and under his influence the *Post* published a poem a day for some years. Mary Willis Shelburne was one of those whose work was published in this context. She visited the family house several times a year in the 1950's, and engaged in conversation with my grandmother, Sarah O'Bryan Frieseke, as well as with the rest of the family. The idea for the correspondence may have been hers, but much of the impulse came from my grandmother, whose impulses

were not safely disregarded. I would have been 13 in 1954. Noelie is my twin sister, Martin younger by a year and a half, Anne older by a year and a half, and Hugh, older still, was finishing high school and on the brink of the seminary, which he attended for a number of years before leaving and becoming a teacher. Rosamond, Matthew and Miriam were progressively younger than I (but Miriam must tell you how old she was in 1954.)

Lewis was absurdly generous in his responses to our letters. Our letters, assembled with the organization of our grandmother, were for us (or for me anyway) a surprisingly familiar way to touch the real character of a person of substance far away. We could not believe then, and I still cannot believe, with what care he read and answered our letters, and how successfully he labored to find something in them to respond to. It seems to me that Lewis's correspondence with Hugh, which Hugh did his best to bring into the theological realm, might show Lewis at his most generous.

There were two further children—Deborah and Jonathan, to bring the entire number to ten.[8]

[AUTOBIOGRAPHICAL ALLUSION(S).]

[1]See LUCY, n. 1.

[2]*VII: An Anglo-American Literary Review,* vol. 20 (2003): 5. See also *Hooper,* 758.

[3]She writes of the experience in *Hooper,* 758–759.

[4]*Hooper,* 759–760.

[5]*Hooper,* 760. Douglas Gresham tells this whole story in his book *Lenten Lands: My Childhood with Joy Davidman and C. S. Lewis* (New York: Macmillan, 1988).

[6]*Letters to an American Lady.*

[7]Many of these letters are found in *Letters to Children,* beginning on 38.

[8]From an e-mail message to the author, November 21, 2004.

"DEEP DOWN INSIDE" — See RIGHT AND WRONG.

DEEPER MAGIC — See MAGIC.

DEEP MAGIC — See MAGIC.

DEPLORABLE WORD, THE[1] — A MAGIC word that has the power to destroy all but its speaker. It is a word that Jadis has learned at great personal cost,[2] and which the great KINGS of CHARN have long known (MN 5). Jadis herself used it to defeat her sister, even though the innocent people of Charn were destroyed as a result.[3] Near the end of MN (15), ASLAN tells the CHILDREN that the people of their world may soon discover a secret as terrible as the Deplorable Word, a broad hint that there are wicked people with the power to destroy all life on earth, most likely a reference to the then newly discovered atom bomb.

[1]This name evokes Percival's Dolorous Stroke of the Arthurian legend, another fateful deed with disastrous consequences (reference given by Dr. David C. Downing).

[2]An echo of ANDREW Ketterley's claim to have learned the secret of Mrs. LEFAY's box at great personal cost, and a warning about the dangers of involvement in magic.

[3]Another echo of Uncle Andrew, this time about his claim that it is permissible to kill the guinea pigs, if necessary, because he owns them. See PAIN.

DEPRAVITY — In the context of discussing the extent to which the WHITE WITCH is human (LWW 8), Mr.

BEAVER observes that though "there may be two views about Humans," there is no question about the witch's EVIL nature—because she is not human but only looks like she is; she is "bad all through." Another way of putting this question is to ask whether the witch is totally depraved, that is, so aligned with evil that she is incapable of being won over to the good. Mr. and Mrs. BEAVER agree that she is entirely evil.

The question remains: How evil are humans? When Mr. Beaver suggests that there are two views on this matter, he is raising one of the major issues of the Protestant Reformation. On the one hand, John Calvin, speaking for the Protestant tradition at this point, believed that humankind has no grace in any of its faculties, that is, that its intellect and will are so aligned with evil that in no sense can people attain lasting happiness unless they repent and attach themselves by FAITH to that perfect man, who is also God: Jesus Christ. Later theologians in the Calvinist tradition understood total depravity to mean not only that humankind is utterly incapable of attaining God on its own, but also that humankind is as bad as it can possibly be. This is one view held of humans.

The other view is expressed by the Catholic theological tradition. Agreeing with the Protestants that humankind, apart from the grace of repentant faith, does not attain lasting JOY, the Catholic tradition distinguishes among the various human faculties, saying of the will that though it is entirely incapable of choosing the God of biblical revelation, it is nevertheless able to develop natural (but not *saving*) virtues. And about the intellect, the Catholic tradition maintains that though the mind remains opaque to the truths to be derived from revelation only, nevertheless it is capable of discerning the maker of the universe from the universe he has made. This latter capability is called "natural theology" by theologians in the Catholic tradition. This is a second view held of humans.

In the characters of ANDREW Ketterley and EMETH, one can discern that Lewis stood on the second, or more Catholic, side of this particular question. Uncle Andrew is at first capable of hearing ASLAN's song, and of recognizing it to be a song. But he chooses to reject this knowledge because the song stirs up in him unwanted thoughts and FEELINGS. By the time it is revealed that the Lion is the singer, Andrew has irreversibly convinced himself that the singing is really roaring (MN 10). On the other hand, Emeth, though he has been consciously seeking TASH all of his life, has really been serving Aslan (LB 15). Emeth has a naturally Christian soul and has only to be shown the truth of his situation for him to immediately acknowledge his allegiance to Aslan.[1]

[1]For more about this question, see UNIVERSALISM.

DESTRIER — CASPIAN'S HORSE, a DUMB BEAST, on whom the prince flees his Uncle MIRAZ (PC 5). It bolts during a thunderstorm and returns to the stables at the castle, thereby unwittingly betraying the fact that Caspian has escaped. *Destrier* is the old French word for "warhorse," and its use by Lewis is another instance of the way in which he creates a medieval atmosphere in the *Chronicles.*

DIGGLE — A DWARF, spokesman for the group of renegade dwarfs who survive the LAST BATTLE, only to be thrown into the STABLE (LB 13). He can't see anything in the stable and assumes that the SEVEN FRIENDS OF NARNIA can't see either; he takes TIRIAN's talk of ASLAN as another attempt to lie to the Dwarfs. The truth is that Diggle sees only what he wants to see. Tirian tells him that the black hole exists only in his IMAGINATION, but when Tirian swings Diggle out of the circle of Dwarfs, Diggle feels like he's been smashed against the wall of the stable.

DIGORY KIRKE — The very wise professor in LWW who welcomes the Pevensie CHILDREN into his large country home when they are evacuated from London in 1940. He is remarkably understanding of youngsters for a fifty-two-year-old, unmarried ADULT. He is later discovered to be the adventurous, curious Digory of MN and the handsome Lord Digory, one of the SEVEN FRIENDS OF NARNIA, in LB.[1] In his anxious concern for his seriously ill mother, he is almost the mirror of Lewis himself, who lost his mother to cancer when he was not quite ten years old.

In the scene where PETER and SUSAN tell the professor about their concern for their sister, the professor uses the Socratic method[2] of probing questions to disclose to them that the best evidence they have points to LUCY's essential truthfulness and sanity, and that they should withhold judgment on the existence of other worlds until they have more evidence. Like many adults, he repeats himself; but unlike most adults, the professor is a courteous listener and hears the children out whenever they ask to speak with him.[3] He also helps them understand that they can't expect to have the same FEELINGS and ADVENTURES over and over again but that they should hold themselves open to experience new feelings and adventures as they come. When he echoes ASLAN's coronation acclamation ("Once a KING in NARNIA, always a king in Narnia"), he ratifies their experience of a world they can reach only by FAITH in good MAGIC.

[1]When Lewis was writing LWW, he did not yet know that the unnamed professor and Digory Kirke were the same character. The professor is modeled both on Lewis's tutor, W. T. Kirkpatrick, a rigorous logician, and on Lewis himself: He had a great sympathy for children who imagined other worlds, and he opened his own home in Oxford to many children fleeing the London blitz at the beginning of World War II. Lewis found in Digory a way to understand his own disappointed HOPES for his mother's return to health. It is not clear from what is now known of Lewis's life if Digory's insatiable CURIOSITY is something Lewis remembered of himself, or if Digory is only a character through whom Lewis can reveal his beliefs about the limits of knowledge and other important themes in MN. See AUTOBIOGRAPHICAL ALLUSION(S).

[2]See PLATO.

[3]Lewis thought that one of the greatest discourtesies a parent can do a child is be dogmatic, interrupting, contradictory, and ridiculing of the things the young take seriously. See *The Four Loves,* 66. See also COURTESY and YOUTH.

DINOSAURS — See DRAGONS, DINOSAURS.

"DLF" — Abbreviation in PC 8, 9, 11, 12, and 13 of "Dear Little Friend," the nickname[1] given TRUMPKIN by EDMUND. Used by all four CHILDREN for such a long TIME that they almost forget what it means.

[1]The use of initials as nicknames is a distinct custom among British schoolboys.

DOG-FOX — The oldest TALKING BEAST present at the party in the woods, which the White Witch breaks up (LWW 11). A dog-fox is a male fox.

DOG(S) — Because he considered dogs "honest, humble persons" (in contrast to CATS), Lewis gave them a large part in the *Chronicles.* Their speech reflects doggy sounds: "How, how? We'll help!" The sheepdog does its job of shepherding by organizing the newly liberated Narnians at the castle of the White Witch for their march to battle, and a hound gets the scent (LWW 16). The dogs use their teeth in battle with the witch's forces. At the CREATION OF NARNIA, a Bulldog is among those who try to decide what form of creature Uncle ANDREW is (10 and 11), and in so doing has an interesting discussion with the She-Elephant about SMELLS and noses. True to his stubborn nature, he objects strongly on three different occasions. In LB 2, a score of talking dogs are sent for by TIRIAN. Later (10), dogs are excited about the prospect of seeing ASLAN. All fifteen talking dogs rally to Tirian's side in the LAST BATTLE (11), and they are among those who pass into ASLAN'S COUNTRY (13). They are very excited about their new life there, and all the new smells. They track down EMETH (14) and take exception to his deprecating use of the term "dog" (they call their misbehaving puppies "boys and girls" [15]).

DOMESTICITY — The *Chronicles* are filled with the delights of HOSPITALITY and domesticity—everything cozy, homespun, handmade, practical, warm, and welcome. Elsewhere,[1] Lewis says that *The Wind in the Willows* is not escapism; rather, it makes us more fit to deal with the harshness of life because "the happiness which this kind of story presents to us is in fact full of the simplest and most attainable things—food, SLEEP, exercise, FRIENDSHIP, the face of nature, even (in a sense) religion." In LWW 7, Mr. and Mrs. BEAVER enjoy quite a comfy domestic scene, of which the CHILDREN are part. In HHB 12, the scene at the home of the three DWARF brothers is almost a paean to domesticity. Through SHASTA, Lewis contrasts CALORMENE meals and homes to Narnian meals and homes, and quite prefers the coziness and homespun, handmade—clearly British—home of the Red Dwarfs. In LB 8, Lewis calls the picture of TIRIAN, EUSTACE, and JILL scrubbing the Calormene pigment off their faces a "homely sight." Jill longs for NARNIA to have the "good, ordinary times" that JEWEL recounts in his "HISTORY" of Narnia,[2] which is filled with DANCES and feasts. Even ASLAN'S COUNTRY does not exclude these homely delights: In Tirian's reunion with his father, ERLIAN, he remembers the "very SMELL of the bread-and-milk he used to have for supper" (LB 16).[3]

[1]C. S. Lewis, "On Stories," *Of Other Worlds*, 14.

[2]For Lewis's love of the ordinary, see LIST(S).

[3]See AUTOBIOGRAPHICAL ALLUSION(S), n. 1.

DONKEY(S)[1] — The first donkey to be created (MN 11) participates in the ANIMALS' discussion about whether Uncle ANDREW is animal, vegetable, or mineral (he thinks if Uncle Andrew is a TREE, then he is quite withered). In HHB 15, ASLAN transforms the unrepentant RABADASH into a donkey. PUZZLE the donkey is the unwilling false Aslan of LB. The donkey is a hieroglyph of the stubborn, foolish, braying person. Puzzle is foolish but he could stand to be more stubborn in his dealings with SHIFT.

[1]*Equus asinus,* the domestic ass, related to the HORSE. The term "donkey" (with its plural) is used thirty-eight times in the *Chronicles,* virtually interchangeably with "ass" (used seventeen times), mostly in its primary sense (referring to the animal) but sometimes in its secondary sense (as a mild insult). "Mules" is used only twice.

DOOR(S) — In the *Chronicles,* doors are often used to make TRANSITIONS between entire worlds. [SPOILERS] In LWW 1, 5, and 17, the Pevensies enter and leave Narnia through the door of the WARDROBE. In PC 15, the CHILDREN and the TELMARINES return through a door in the air, and in SC 1 and 16, the door of the garden wall of EXPERIMENT HOUSE leads to Narnia and ASLAN'S COUNTRY. In VDT 16, EUSTACE, EDMUND, and LUCY return from Narnia to ENGLAND through another door in the air. In LB 12, 13, 14, and 16, the door is an image of DEATH and ESCHATOLOGY. The STABLE door is viewed darkly by POGGIN and TIRIAN and JILL, but with HOPE by JEWEL, who perceives that it may be the door to Aslan's country. This door is free-standing; it can be walked around and looked at, but when looked through, Old Narnia can be seen. Looked *at*, the door leads "from nowhere to nowhere"; looked *through,* one can see the scene in front of the stable. The door[1] flies open at Aslan's command, and good creatures begin to pour through into Aslan's country. At the END OF NARNIA, when the sea rises, it never rises high enough to pass through the threshold of the door, and when the GIANT* TIME snuffs out the SUN, PETER solemnly locks the door with his golden KEY.

[1]Beginning in LB 13 and concluding with 16, Lewis refers to "Door" and "Doorway" nineteen times, capitalizing them when he intends to signify the boundary between time and eternity, and leaving them lowercase to signify the death of an individual.

DRAGON ISLAND — A country of mountainous cliffs and crags encircling a bay, like a Norwegian fjord (VDT 5). The island is beautiful to look at but inhospitable. It is here that Eustace finds a dead DRAGON—probably the one who devoured Lord OCTESIAN—and becomes a dragon himself. It is also the scene of EUSTACE's conversion. From his dragon's-flight perspective, Eustace sees the island to be entirely mountainous and populated only by wild goats and swine.

DRAGONS, DINOSAURS — Dragons, the fire-breathing giant lizards of world MYTHOLOGY, are certainly related to the giant lizards of prehistoric times. From the Greek word *drakon,* which means "to watch," dragons are usually guardians of some sort. The dying dragon that Eustace encounters is guarding its pile of treasure (VDT 6). EDMUND comments later that all dragons collect gold, and indeed it is in stealing some of the golden treasure for himself that EUSTACE, in his GREED, is transformed into a dragon. Lewis's dragon has a long snout the color of lead, claws made for tearing, a long tail, and batlike wings. Its favorite food, when it can get it, is fresh dragon. In LWW 9, a stone dragon is one of the statues in the courtyard of the Castle of the White Witch. In UNDERLAND, JILL, Eustace, and PUDDLEGLUM see dozens of sleeping dragon- and batlike creatures (SC 10); at the END OF NARNIA (LB 14), these same beasts awake and go about clearing all life from

Eustace weeps huge tears when he sees his dragonish reflection. He wishes more than anything to be human again and back with his companions. (VDT 6)

NARNIA. They grow old and die and their skeletons dot the dead countryside.

DREAM(S) — Dreams, the blessing and bane of SLEEP, play a large part in the *Chronicles.* CASPIAN, who misses his NURSE and hates life in MIRAZ's castle, dreams of Old Narnia to console himself. In SC, JILL has anxious dreams in which she works out her FEAR over the possible failure of her QUEST. [SC SPOILER] She wants to believe ASLAN is a dream so that she won't have to keep going, and her dream (which she does not consciously remember) in which the wooden HORSE turns into Aslan is perhaps not really a dream at all, but Aslan's way of encouraging her to remember the SIGNS and to continue. Life sometimes seems like a dream, and the QUEEN OF UNDERLAND tries to make the three travelers think that their previous experience in Narnia was a dream (ironically, after leaving UNDERLAND, they think *it* was a dream). In MN, Strawberry compares his vague remembrance of his previous existence as a cab-horse to a dream, and when HELEN first enters Narnia she thinks she is dreaming. TIRIAN's vision of the SEVEN FRIENDS OF NARNIA is like a dream in which he cannot make himself heard.

It is in VDT [SPOILER], in the experience of the DARK ISLAND, that the subject of dreams becomes most vivid, and Lewis considered this so important that he made several substantial changes in the text between the British and old American editions of the book.[1] After the *DAWN TREADER* emerges from the darkness, the British edition (now the English edition for the world) says, "all at once everybody realized that there was nothing to be afraid of and never had been. They blinked their eyes and looked about them." In the old American editions, Lewis deletes these two sentences entirely, thinking perhaps that he was making too little of the reality of which the travelers were afraid. He replaces these sentences with one long, beautiful simile, one of the finest he ever wrote:

> And just as there are moments when simply to lie in bed and see the daylight pouring through your window and to hear the cheerful voice of an early postman or milkman down below and to realise that *it was only a dream: it wasn't real,* is so heavenly that it was very nearly worth having the nightmare in order to have the joy of waking; so they all felt when they came out of the dark.

This is a major change: Lewis here is highlighting the JOY of waking after a night of fear.

The next change comes by way of an omission. Both editions print the expectation the crew has that the ship will be covered with grime and scum. The British edition goes on to say: "And then first one, and then another, began laughing. 'I reckon we've made pretty good fools of ourselves,' said Rynelf." The old American editions delete both sentences, thereby removing another denigration of the seriousness of night fears.

When Caspian asks what boon Lord RHOOP wishes the KING to grant, the British edition prints:

> 'Never to bring me back there,' he said. He pointed astern. They all looked. But they saw only bright blue

sea and bright blue sky. The Dark Island and the darkness had disappeared forever.

'Why!' cried Lord Rhoop. 'You have destroyed it!'

'I don't think it was us,' said Lucy.

Lewis reconstructs this entirely for the old American editions:

"Never to ask me, nor to let any other ask me, what I have seen during my years on the Dark Island."

"And easy boon, my Lord," answered Caspian, and added with a shudder, "*Ask* you: I should think not. I would give all my treasure *not* to hear it."

This is perhaps the greatest difference between the editions. The British edition says that our Dark Islands in life can be destroyed; the old American editions are much more real in their assessments.

Finally, the old American editions, having deleted the destruction of the Dark Island, add a parting note about the experience. All editions print the sentence: "So all afternoon with great joy they sailed south-east with a fair wind." To this the old American editions add the independent clause: "and the hump of darkness grew smaller and smaller astern."

[1]In 1994 the editors at HarperCollins*Publishers* decided to make the British edition the standard. Between the time Lewis corrected the galleys of the British edition and the time he corrected those for the first American edition, he delivered his address "On Three Ways of Writing for Children." In speaking to the objection that fairy tales are too violent for children, Lewis said: "I suffered too much from nightfears myself in childhood to undervalue this objection. I would not wish to heat the

fires of that private hell for any child" (*Of Other Worlds*, 30). See AUTOBIOGRAPHICAL ALLUSION(S), VIOLENCE, and *Letters to Children*, 33–34.

DRINIAN — A lord of NARNIA and captain of the *DAWN TREADER*. The dark-haired Drinian is a straightforward man, a respected seaman, and a good ADULT. [VDT SPOILERS] A stickler for fairness, he discusses with EDMUND whether EUSTACE should be handicapped in some way to compensate for his greater size in his fight with REEPICHEEP. When he sets sail on a southeast course, he lets everyone not on duty rest. And like CASPIAN, he hates the SLAVE TRADE; he wants to board the slave ship and retake the captives. A good sailor, he rigs a jury-mast (a temporary mast) in order to have some sail after the mainmast is brought down in the storm. CORIAKIN uses Drinian's knowledge of GEOGRAPHY to make two MAGIC maps. The captain's advice is always practical: He advises Caspian not to proceed into the darkness and asks Reepicheep what practical use it would be to go farther. A good leader, he does not hesitate to follow the ALBATROSS. In SC 4, Drinian is still in the court of King Caspian. He has become Prince RILIAN's chief older friend and has advised him to give up his QUEST for the serpent. He is not taken in by the QUEEN OF UNDERLAND, recognizing her at once as EVIL. Drinian is among those summoned to the GREAT REUNION in LB.

DRYAD(S) — In Greek and Roman MYTHOLOGY, wood NYMPHS who live and die with the TREES of which they are the spirits. Their creation by ASLAN is not specifically noted, but they appear to have come into being in response to Aslan's command for NARNIA to awake. They are also part of the wild WOOD PEOPLE. Dryads like to DANCE and are mentioned as dancing partners of FAUNS. Physically they reflect the characteristics of their trees; for example, birch-girls are pale and head-tossing, and the beeches are regal. Perhaps the most poignant passage concerning Dryads is in LB 2 [SPOILER], when a beech dryad announces that forty of her brothers and sisters have been killed in the name of the false Aslan. As her tree is cut down, miles away, she, too, dies.

Throughout the *Chronicles,* oaks are male. With the birch-girls and willow-women are oak-men and holly-men (who have wives, bright with berries). See LWW 2, 9, 12, 13, and 16; PC 4, 6, 9, 10, 11, 14, and 15; VDT 2; HHB 5; SC 3; and MN 10 and 15.

DUFFER(S) — These simple, childish subjects of CORIAKIN, also known as Monopods (*mono,* "one" + *pod,* "foot") and Dufflepuds, are invisible when first introduced in VDT (9, 11, and 14). The Duffers (once common little DWARFS) have a single center leg, three feet long, which ends in a large foot that resembles a canoe. They sleep with the foot raised in the air like an umbrella and curl up under it, looking very much like large toadstools. It is a shape Coriakin gave them as punishment when they refused to do what he asked. Thinking this shape ugly, they had the CHIEF VOICE's young daughter CLIPSIE recite a spell to make them invisible. They echo every word of their silly leader. But according to Coriakin, it is better that they admire such a leader than admire no one at all. Their foolishness and immaturity show up in many ways: They do not know enough to grow their food, so Coriakin must order them to grow a garden; they water it from a spring rather than from the stream that flows from the spring right by the garden; they wash up the plates before dinner to save time afterward; and they plant boiled potatoes so that they do not have to cook them afterward. [VDT SPOILERS] By the time the *DAWN TREADER* lands on the island, the land of the Duffers, they are tired of their invisibility and threaten to harm Lucy and her companions if she does not read the visibility spell from the MAGICIAN's book (a young girl is the only one who can work the charm). The visible Duffers are a delight to watch: They bounce along on their single feet, cheering for LUCY and agreeing with her opinion that they are not ugly even as they also agree with Chief Voice's opinion that they are. When REEPICHEEP addresses them as Monopods, they relish the name but cannot get it straight, corrupting it to Moneypuds, Pomonods, and Poddymons. Finally, mixing it up with Duffers, they arrive at the name of Dufflepuds, which Lewis says they will probably be called for centuries (even though NARNIA has little more than 140 years to go). Discovering, at Reepicheep's suggestion, that their feet make them very buoyant, they take great pleasure and pride in sailing about the harbor, much to the delight of the adventurers who teach them to race and play games. Only the Chief Voice, perhaps because his authority is threatened, refuses to participate in the fun, warning his fellow Duffers that the water is "powerful wet." As the *Dawn Treader* moves to the north of the harbor, the Duffers accompany it with cheers, and RYNELF coins a simile for his shipmates, "as silly as those Dufflepuds."

DUFFLE — The Red Dwarf who rescues Shasta from the indecisive animals and takes charge (HHB 12). A practical person, he notices Shasta's hunger, mutters a reproach against himself and his neighbors for their inhospitality, and hustles the boy off to a breakfast with Rogin and Bricklethumb, Duffle's brothers. Their home life is the very icon of the loveliness of the ordinary and of domesticity. Any meaning their names might have (*duffle* is a kind of coarse cloth, and *brickle* is an adjective meaning "fragile") is secondary to the quaint, friendly sound the names have.

DUFFLEPUD(S) — See Duffer(s).

DUMB BEAST(S) — The animals of Narnia who do not have the gift of speech, and who are given by Aslan into the care of the Talking Beasts. Dumb and witless animals are smaller than Talking Beasts—indeed, they are the size of normal animals—and they are always grave. Though like all Narnians they are to be respected, it is permitted to kill them for food. Thus Eustace, Jill, and Puddleglum must hunt birds for food, and a hunter has killed the dumb lion whose skin is worn by Puzzle. In instructing the Talking Beasts on their responsibilities, Aslan tells them not to go back to their dumb ways.[1] In LB, the reduction of Ginger to dumbness awakens in the Talking Beasts their "greatest terror": that any beast who is not good will be struck dumb.

[1]In PC 9 the Pevensie children and Trumpkin are stalked by a wild bear, who is killed by the Dwarf just in the nick of time. At this point in Narnian history, so long past the Golden Age, many Talking Beasts have gone dumb and it is often impossible to tell the two apart. Lucy ponders how awful it would be if

human beings could become wild inside but still look the same on the outside. The ever-practical Susan cuts her off, saying they have enough problems as it is. The implication is that this has already happened in our own world. See also the LAPSED BEAR OF STORMNESS.

DWARF(S) — One of the NINE CLASSES OF NARNIAN CREATURES listed by Doctor CORNELIUS (and common in world MYTHOLOGY), this race of rational beings (human enough to intermarry with humans—see CORNELIUS, DOCTOR) comes into existence at ASLAN's command for NARNIA to awake at the CREATION OF NARNIA in MN, figures prominently in PC and LWW (but scarcely in VDT, SC, and HHB), and (after its ranks are thinned by Aslan's JUDGMENT) passes into ASLAN'S COUNTRY after the END OF NARNIA.

There are three kinds of Dwarfs in Narnia: Red Dwarfs, Black Dwarfs, and DUFFERS. The Duffers are the smallest and least distinguished of the three; the Red and Black Dwarfs are three to four feet tall, deep-chested, and stocky. All species are bearded, the Red Dwarfs with hair like foxes' and the Black Dwarfs with hard, thick, dark, HORSElike hair. With the exception of the Duffers, Narnian Dwarfs are miners, smiths, and metal craftsmen. They excel as tailors and are noted woodsmen. In war, they are fierce axe-wielders and deadly archers, marching and communicating by the SOUND of drums. In peace, Red Dwarfs are treasure-seekers and love wild DANCES and rich feasts and brightly colored clothes (the peacetime activities of the Black Dwarfs are not described, but it is presumed that they are hard workers, though not given to REVELRY). No Dwarf wives are mentioned but one of the results of the ending of the Hundred Years of Winter is that young Dwarfs will not have to go to SCHOOL.

The chief difference between the two main species of Dwarfs is seen in their reaction to Aslan in PC: At their first sight of the Lion, the Red Dwarfs are open-mouthed and speechless though they know he has come as a friend, but the Black Dwarfs begin "to edge away." TRUMPKIN, Rogin, Bricklethumb, DUFFLE, and the SEVEN BROTHERS OF THE SHUDDERING WOOD are the best examples of Red Dwarfs. Only groups of Red Dwarfs are called brothers; Black Dwarfs, though they may live together, don't seem to have fraternal feelings. NIKABRIK, GRIFFLE, DIGGLE, and the WHITE WITCH's driver are clearly Black Dwarfs (see also the FIVE BLACK DWARFS). From their feistiness and dry wit, it would appear that the two Dwarfs in the Narnian embassy to TASHBAAN in HHB are Red. It is not at all clear to what species POGGIN belongs; and since at least one of the renegade Dwarfs in the LAST BATTLE does come into Aslan's country through the DOOR of judgment (this one Dwarf is not to be confused with the group of Dwarfs *thrown* through the door by thc CALORMENE soldiers), we cannot therefore assume that Poggin or these are Red.

The TALKING BEASTS have their own opinions about the Dwarfs. Mrs. BEAVER says she knows good Dwarfs, but her husband, admitting that he knows some good ones too, says that their numbers are very few and these are the least like MEN. TRUFFLEHUNTER finds that Dwarfs are as forgetful and changeable as humans. He adds, in defense of Doctor Cornelius, that Dwarfs and half-Dwarfs are much less different than are Talking Beasts from DUMB BEASTS.

Concerning half-Dwarfs, Doctor Cornelius is the only named example; CASPIAN'S NURSE may also have Dwarf blood. During the years of TELMARINE oppression, the Dwarfs tried to disguise their race by, in some instances, intermarrying with the conquering humans. The Black Dwarfs consider the half-Dwarf traitors worthy of DEATH.

Caspian and his son, RILIAN, are especially deferential to Dwarfs; the former names Trumpkin his regent and the latter, upon his release from UNDERLAND and his reception of the creatures' act of OBEDIENCE, addresses the oldest of the Dwarfs as "father" and seeks his counsel. This mirrors Aslan's own respect when he chooses a Dwarf to be a member of the first council (for protecting Narnia against Jadis). Aslan calls upon them to fashion crowns for King FRANK and Queen HELEN, and Dwarfs are the king's train-bearers.

[LB SPOILER] The story of the treachery of the renegade Dwarfs in LB is one of the saddest in the *Chronicles.* Manifesting the suspicion their race is prone to and their tendency to follow a leader doggedly (witness the Duffers and their CHIEF VOICE), the Dwarfs echo Griffle's declaration of independence and take up a refrain that is chanted five times in the last half of LB: "The Dwarfs are for the Dwarfs." This harks back to Nikabrik's continuous complaint that his forces were being sacrificed for the protection of Caspian's army in the WAR OF DELIVERANCE. Nikabrik refused to be "taken in," and his ancestors are condemned to repeat history. Aslan judges that they have rendered themselves incapable of help, chosen cunning instead of FAITH, and locked themselves into the prison of their minds and are, therefore, in the FEAR of being taken in, never to be taken out. The renegade Dwarfs are last seen "crowded together in their imaginary STABLE." Nothing further is known of their fate.

E

EARTHMAN, EARTHMEN — Inhabitants of UNDERLAND and natives of Bism (SC 10, 11, 13, and 14). These pale gnomes[1] come in all shapes and sizes, from less than a foot high to over seven feet tall, and have all sorts of faces—one even has a horn in the middle of its[2] head. They are all profoundly sad and SILENT. Perennially busy, they move in closely packed groups with padding, shuffling steps. The WARDEN OF THE MARCHES OF UNDERLAND leads a group of one hundred Earthmen armed with three-pronged spears. [SC SPOILER] They do the work of the QUEEN OF UNDERLAND and have been digging a tunnel to within only a few feet of the Overworld, through which she and RILIAN intend to conquer the updwellers. When the queen has been killed and the enchantments lifted, the Earthmen are freed from her spell to become their true selves—fun-loving, jig-dancing, somersaulting, firecracker-firing people who shout in loud voices and dart all over the place. GOLG reveals that they were taken from Bism, deep inside the earth, to become slaves in Underland—what they call the Shallow Lands. Once they remember who they really are, they rejoice that the witch is dead and plan to return home.

[1]According to Paracelsus, gnomes inhabit the element earth. For a more detailed discussion, see SALAMANDERS OF BISM.

[2]Although they are named "Earth*men*," the impersonal pronoun "it" is used by Lewis throughout all appearances or discussions of these gnomes.

EARTHQUAKE(S) — Portents of the coming of the END OF NARNIA, especially in LB. When SHIFT professes his disbelief, the earth shakes (1), an event PUZZLE rightly interprets as a SIGN that he should not be wearing the lion's skin. Shift, on the other hand, interprets it to be a confirmation of his plan. Another earthquake is felt at Shift's destruction (11), and again when RISHDA is thrown into the STABLE (12). The earth trembles as ASLAN approaches the DWARFS (13) and Aslan growls so mightily at EMETH's restatement of Shift's lie (that Aslan and TASH are one) that the earth shakes (15). In MN 4, the SOUND of the GOLDEN BELL intensifies until an earthquake destroys the room in which DIGORY and POLLY stand.

[ASLAN'S VOICE; BIBLICAL ALLUSION(S).]

ECOLOGY — Ecology is the harmony of nature, and the disruption of this harmony in NARNIA is always a SIGN that something is fundamentally wrong. As EDMUND is wandering through the forest alone, having already fallen under the spell of the WHITE WITCH's TURKISH DELIGHT, he daydreams about what he will do when he is KING:[1] make decent roads, inaugurate railway lines, and make laws against beavers and dams. In MN, Uncle ANDREW thinks along the same lines: Impressed by the fertility of the newly created Narnia, he can hardly wait to begin commercial exploitation of its resources. The TELMARINES' rule is especially marked by disregard for nature: They have felled TREES, defiled the streams, and forced the DRYADS and NAIADS—the very souls and spirits of nature—to retreat into a deep SLEEP. The bridge at the Fords of Beruna is also seen as a human imposition on nature, and the RIVER-GOD calls it his "chains." At ASLAN's command, it is destroyed and reclaimed by nature through the powers of BACCHUS. In LB, the degeneration of society is highlighted by the wholesale felling of talking trees and the mining of ore by slave labor.

[1]See TECHNOLOGY.

EDMUND PEVENSIE — The second son and third CHILD of Mr. and Mrs. PEVENSIE, first a traitor and later King Edmund the Just in the GOLDEN AGE OF NARNIA, who grows from the sensual, difficult, jealous nine- or ten-year-old in LWW to the handsome and brave twenty-four-year-old KING of HHB and the helpful and playful nineteen-year-old youth who is mortally hurt in the RAILWAY ACCIDENT in LB.

The reader joins Edmund's STORY when the Second World War separates the boy from his parents. He has been attending a bad SCHOOL, which has had a negative effect on him. He is struggling with his older brother, PETER, who is always considered better than he, and his older sister, SUSAN, who has taken upon herself the role of his mother. The only person he ranks above is his younger sister, LUCY.

Even from the ANIMALS Edmund gets excited about, the reader senses that the boy might be the "bad apple in the barrel."[1] Edmund is seduced by the EVIL* MAGIC in the TURKISH DELIGHT the WHITE WITCH offers him. He lies about his and Lucy's experience of NARNIA, letting her down terribly.

Lewis is careful to distinguish Edmund's motives when he decides to betray his siblings to the witch: He doesn't want any real harm to come to his brother and sisters but (besides power and pleasure) he does want revenge on Peter, and for the witch to prefer him to the others. But "deep down inside"[2] he really knows how cruel and evil she is. As he trudges across the frozen landscape to her castle, he consoles himself with the thoughts of how he can use TECHNOLOGY to transform Narnia into what he considers a better place.

It is only when the witch shows her real cruelty on their journey to the STONE TABLE that Edmund begins to experience a change of heart: He wants to be a loving and beloved brother again. When the witch petrifies the celebrating Narnians, Edmund's conversion manifests itself in full force: "FOR THE FIRST TIME" he feels sorry for someone other than himself. The spring that is bursting forth all around him in Narnia is also happening inside him.

After Aslan's forces free Edmund, he and the Lion have a private conversation and then he is reintroduced to his siblings. And though the witch calls him a traitor, Edmund keeps his eyes fixed on Aslan[3] and instinctively knows that he is expected only to wait and obey. He breaks the witch's WAND in the first part of the battle but suffers a wound thereafter. When Lucy heals him with her CORDIAL, not only does he recover from the injury, but he is also rejuvenated as a person.

At the beginning of PC [SPOILERS], Edmund is excited about not having to return to his studies and about having an experience similar to the ADVENTURES in books he has been reading about people being shipwrecked and finding fresh water and food in the forests. Unlike with his first visit to Narnia, Edmund is delighted to be there; when he is asked to decide whether or not they should follow Lucy, he remembers his betrayal of her in LWW and sides with her. Aslan salutes him with a vigorous "Well done" and breathes on him;[4] an air of greatness henceforth hangs on him, something apparent even to GLOZELLE and SOPESPIAN, MIRAZ's treacherous lords. His nobility is also apparent in his genuine sorrow over the fact that Peter and Susan aren't going to be allowed to return to Narnia again.

In HHB 12, King Edmund is the first Narnian ADULT that SHASTA sees and the first adult that Shasta admires. When Shasta is later introduced to Edmund, the king assures the boy that he is not a traitor for overhearing their plans—Edmund knows what betrayal is really about.

In the course of VDT Edmund comes to appreciate his cousin EUSTACE as one who has been SAVED by Aslan from being a really terrible person. The scientist in Edmund sets Lucy straight when she confuses the MERMEN and mermaids with the SEA-PEOPLE. And he discusses the difference between a flat world and a round world with Caspian. When Caspian wants to abdicate, Edmund compares Caspian's proposal to Ulysses' loss of reason under the influence of the song of the Sirens in Homer's *Odyssey*.

Edmund shares with Lucy and Eustace the vision of the LAMB at WORLD'S END. It is in answer to Edmund's question that Aslan reveals Lewis's intention for writing the *Chronicles*. And at the very end of LB Edmund recognizes in Aslan the one whom he has gotten to know by his earthly name.

[1]It must be presumed that Lewis made the change in the pre-1994 American editions of LWW 1 that replaces the *foxes* that the "British" Edmund is excited about with the *snakes* of the "American" Edmund. (The "British" Susan prefers *rabbits,* while the "American" Susan enthuses about *foxes,* perhaps a veiled reference to her desire for a high social life, riding to the hounds, and the like.) The HarperCollins editions restore the British originals.

[2]See RIGHT AND WRONG.

[3]See ASLAN and BIBLICAL ALLUSION(S).

[4]See ASLAN'S BREATH.

EFREETS — EVIL spirits, allied with the WHITE WITCH and present at the slaying of ASLAN (LWW 14). The word is a variant of *afreet,* which denotes an evil demon or monster of Islamic MYTHOLOGY. They are often associated with GHOULS.

EMETH — The noble CALORMENE* TARKAAN who yearns to serve TASH and is embraced by—and ultimately embraces—ASLAN (LB 10, 13, and 14). *Emeth* is the Hebrew word for "faithful, true," a deliberate reference by Lewis. This junior officer is a darkly handsome young man who has LONGED since his youth to serve Tash. He wants finally to meet him face-to-face, even if this should mean DEATH; convinced that the real Tash is inside, he goes through the STABLE* DOOR. Finding himself in a sunlit world, he assumes it is Tash's country and goes in search of his GOD. Actually, however, it is ASLAN'S COUNTRY, and he finds

Aslan, whom he recognizes as a great Lion. Emeth claims to be unworthy because he has served Tash all his life, but Aslan says that he will account it as service to himself. Emeth's mystical sense is indicated in his words "It is better to see the Lion and die than to be TISROC of the world and not to have seen him."

[BIBLICAL ALLUSION(S); DEPRAVITY; UNIVERSALISM.]

EMPEROR-BEYOND-THE-SEA[1] — ASLAN's Father, author of the Deep MAGIC and the Deeper Magic. In almost every instance where an explanation of Aslan is given, the formula "the great Lion, the son of the great Emperor-beyond-the-Sea" is used, almost as if it were part of a PROFESSION OF FAITH. The "sea" here is the Eastern Sea, and "beyond" suggests both ASLAN'S COUNTRY and the GREATNESS OF GOD.[2] The Emperor is never pictured, but his SCEPTRE is mentioned. In the remarkable scene of SHASTA walking in the night fog, the Emperor's voice is heard as the first "Myself"[3] pronounced by the Large Voice in deep, low, EARTHQUAKE tones. In ASLAN'S NAME and in the name of the Emperor, one of the seven KINGS and queens, most likely the High King PETER, dismisses TASH to go to his "own place" (that is, HELL).

[1]"Emperor-over-Sea" (printed both with and without hyphens and the definite article) is used interchangeably with "Emperor-beyond-the-Sea" throughout the *Chronicles* in all editions.

[2]There is no mention in the *Chronicles* of where the Emperor might live. Nowhere is he connected with Aslan's country, and we can only guess, on analogy with Christian theology and MYTHOLOGY on this subject, that he is the ultimate goal of the "further up and further in" process. There is, however, one tantalizing clue, found on the last page of VDT, when Aslan tears a

Coriakin the Magician and Lucy can't help laughing at the antics of the Duffers. Visible once more, they are hopping about on their monopods like enormous fleas. (VDT 11)

DOOR open in the sky and Edmund, Eustace, and Lucy have a momentary glimpse of a "terrible white light from beyond the sky."

[3]See TRINITY.

END OF NARNIA — At ASLAN's command, a series of ESCHATOLOGICAL events heralds the end of NARNIA (LB 14). The GIANT* FATHER TIME winds his HORN and the STARS fall from the heavens. The DRAGONS and dinosaurs round up all living things and send them scurrying toward the STABLE* DOOR for an experience of JUDGMENT. Then the dragons and dinosaurs eat all the forests until there is nothing left but dead rock. The sea rises, as does the dying red SUN. The Moon rises in the wrong position, too close to the sun, which absorbs her. The giant squeezes the sun out, and all is ice-cold, total darkness.

ENGLAND — An island nation of western Europe, the birthplace of the SEVEN FRIENDS OF NARNIA and SUSAN. Things Narnian are larger than life when seen against the English sky: Queen Jadis looks more beautiful, fiercer, and wilder in London than she does in CHARN; the Apple of Youth is filled with a glory that causes earthly things to pale in comparison. According to DIGORY, who loves the philosopher PLATO, England is only a shadow of something in ASLAN'S COUNTRY. When PETER, EDMUND, and LUCY see England and Professor Kirke's HOUSE from the GARDEN, TUMNUS explains that they are "looking at the . . . inner England [where] no good thing is destroyed" (LB 16).

ERIMON — One of the great TELMARINE Lords under CASPIAN IX (PC 5). He is executed with Arlian and a dozen other lords under false charges of treason invented by Miraz.

ERLIAN — Father of Tirian; second to the last king of Narnia. Erlian and Tirian are reunited at the great reunion in the real Narnia (LB 16). In Aslan's country, Erlian is transformed into a hearty young adult, the father Tirian remembers from his youth. The charged emotional quality of Erlian's embrace of his son is unique among the parent-child relationships in the *Chronicles*. Tirian's memories in conjunction with the reunion are of childhood pleasures so similar to Lewis's own autobiographical details that it is very likely this scene is Lewis's literary reconciliation with his own father.

ESCHATOLOGY — In Christian theology, the study or science of the *eschata* (Greek for "last things"), what happens at the end of an individual's life and at the end of time. Traditionally the "four last things" are death, judgment, hell, and heaven. To live with hope is to live with the expectation that each of us will die, will be judged, and will enter into heaven. LB begins, "In the last days of Narnia ..." This dying world is marked by the absence of Aslan and with the doubt by many Narnians that he ever existed. Jewel says, "This is the end of all things." Jill expresses the hope that, unlike earth, Narnia will go on forever, but Jewel declares to her that "all worlds draw to an end; except Aslan's own country." Roonwit's dying message is "To remember that all worlds draw to an end and that noble death is a treasure which no one is too poor to buy." Ultimately, the end of Narnia unfolds, and the creatures of Narnia confront Aslan's judgment before they can pass through the DOOR. They must look into Aslan's face; about this they have no choice. Some look at Aslan with hatred, and these disappear into the blackness of his shadow. Others look at Aslan with love (not unmixed with proper FEAR—a NUMINOUS experience), and these are allowed to pass through the door into ASLAN'S COUNTRY.

ETTINS — That they are EVIL* GIANTS is evidenced by the fact that they are members of the WHITE WITCH's party and present at the slaying of ASLAN (LWW 14). *Ettin* is another form of the word *eten* (both words mean "giant").

ETTINSMOOR — The territory of the ETTINS north of the River Shribble and south of HARFANG, near the ruined city. It is mentioned first by GLIMFEATHER as being on EUSTACE and JILL's path to the north (SC 4, 5, and 6). PUDDLEGLUM points out some "hills and bits of cliff" as the beginning of Ettinsmoor and tells Eustace and Jill that GIANTS live there. Countless streams cover the moor. The name is derived from *ettin,* an obsolete form of the word *eten,* which means "giant"; and *moor,*[1] an open expanse of fertile land.

[GEOGRAPHY, NARNIAN.]

[1]In the British edition of LB and in the HarperCollins editions, Lewis spells the word "Ettinsmuir." *Muir* is another form of *moor,* and it is likely that he forgot how he had spelled it in the earlier books.

EUAN, EUAN, EUOI-OI-OI-OI — See BACCHUS, n. 1.

EUSTACE CLARENCE SCRUBB[1] — [VDT and SC SPOILERS] The only CHILD and son of HAROLD and

ALBERTA Scrubb, and a cousin of the Pevensie children. He is a victim of his parents' untraditional ways of child raising and of his schooling at EXPERIMENT HOUSE but is reformed in VDT by being transformed into a DRAGON (the outer form of his inner disposition) and is transformed again by Aslan himself. JILL'S FRIEND and fellow adventurer in SC and one of the SEVEN FRIENDS OF NARNIA in LB, he is one of the most memorable characters created by C. S. Lewis.

Lewis wants the reader to see even in Eustace's first diary entry many clues to the boy's character defects: Eustace magnifies the normal sea conditions into a "frightful storm"; he ignores the fact that he has been healed of his seasickness, boasting that he is not ill. He is not grateful; he thinks himself the only one aware of what he supposes to be their alarming condition; he blames the others' not noticing on either their conceit or their COWARDICE; he is patronizing, arrogant, jealous, and complaining.

His transformation takes place on DRAGON ISLAND, when he wanders away from the group in order to avoid the work of refitting the ship. He tries to make himself comfortable, but he begins to feel lonely "FOR THE FIRST TIME" in his entire life. Becoming a dragon improves Eustace's character because he is concerned about helping his friends. And the pleasure of liking people and being liked is the most powerful antidote to the discouragement he often feels as a dragon. The fact that he is a nuisance as a dragon eats into his mind as painfully as the arm-ring eats into his foreleg. In his PROVIDENCE, Aslan appears to him in a DREAM and cleanses Eustace of his dragon-self. When the boy returns to his companions, he tells EDMUND his STORY.

Eustace begins to be a different, better person from this point on, though with occasional relapses.[2] He is brave "for the first time" when he fights the SEA SERPENT. When his friend REEPICHEEP volunteers to stay overnight at ASLAN'S TABLE, Eustace also decides to stay, something Lewis says is especially brave of him because he hasn't even been able to prepare himself by reading about all the things that could happen during the night. Eustace returns to earth with a touch of the Lion's mane and the Lion's kiss.

The reader sees a very different Eustace at the beginning of SC: He is full of empathy for Jill, who is CRYING over being bullied by the GANG. Eustace wants her to see how he's changed—by standing up to CARTER, who was cruel to ANIMALS, and by defending SPIVVINS. He wants very badly to share his Narnian experiences, so he risks telling Jill about Aslan. He shows how much he really understands Aslan's ways when he does not permit Jill to try to force her way into Narnia with dark MAGIC. He knows that the best way in is to ask.[3] They find themselves in ASLAN'S COUNTRY.

When ASLAN'S BREATH blows Eustace into Narnia, the atmosphere restores him to the strength he had at the end of VDT. When the QUEEN OF UNDERLAND tries to reenchant RILIAN and his three rescuers, Eustace is able to argue clearheadedly against the witch's reductionisms. After the queen is killed, Eustace apologizes to Jill for his bad temper and wishes her well as they climb up to Narnia. From this time forward they call each other by their Christian names.

The friendship of Jill and Eustace that began in SC continues into LB. The LAST BATTLE finds him hoping that he will be brave and performing well in the face of his FEARS. He is captured and thrown through the STABLE* DOOR. Like the other friends of Narnia, he is miraculously refreshed. He is last mentioned as hesitating at the GREAT WATERFALL. He thinks it crazy to swim up the falls. But he tries Lucy's experiment—to feel afraid—and he can't. So up he swims—a far cry from the weak and whiny brat of VDT.

[1]It is conceivable that Lewis is poking fun at his own name, Clive Staples Lewis, and at how unpleasant he thought himself to be when he was a boy only a little older than Eustace. Lewis was known to his friends and family as Jack, a name he chose for himself when he was four.

[2]This is Lewis's picture of how repentance is a process, beginning at a definite moment of conversion, yes, but extending through the rest of a person's life.

[3]Yet another of Lewis's hints at the OBEDIENCE and simplicity required of authentic PRAYER.

EVE — See SON OF ADAM, DAUGHTER OF EVE.

EVIL[1] — Evil stands watching the CREATION OF NARNIA in the person of Jadis, and NARNIA is never completely free from its presence. Jadis, driven out of Narnia in MN, comes back as the WHITE WITCH to bring the Hundred Years of Winter in LWW. And though she is killed by ASLAN, in PC the HAG informs Doctor CORNELIUS that one can't really kill a witch. This is Lewis's way of saying we must always keep a constant watch against evil. During the GOLDEN AGE OF NARNIA, the smaller ANIMALS have become so COMFORTABLE and secure that they fail to recognize the first SIGNS of danger. In MN, Aslan calls the first council to discuss what to do about the evil that has entered Narnia.

The animals, still innocent, mishear the words "an evil" as "a Neevil" and think it is some sort of creature, perhaps Uncle ANDREW. In this they are not far from wrong, for his dealings in dark MAGIC have brought the foolish man into direct contact with evil; he is, in fact, in love with the personification of evil, Jadis. The CHILDREN are called into Narnia each time to help contain evil that has gotten out of hand. In LB, evil has taken over most of Narnia, and its presence is detected in the depression and gloom that hang over that world like a poisonous cloud. Now it is too much even for the SEVEN FRIENDS OF NARNIA to contain, and in the end Narnia is unmade by Aslan, its maker. The creatures who, at the DOOR of JUDGMENT, look on Aslan with hatred are banished eternally; evil is consigned to "its own place," and thus there is no evil in ASLAN'S COUNTRY.

[1]Lewis owes his theology of moral evil to St. Augustine (died C.E. 430). Evil is essentially parasitic (*Mere Christianity,* Book II, Chapter 2, ¶10), as Augustine also thought: "Evil is the privation of the good."

EXPERIMENT HOUSE — The SCHOOL at which JILL and EUSTACE are students. Lewis's view of modern educational methods is made quite clear in his description of Experiment House. The woman principal, the HEAD, is more interested in the bullies than the better-behaved CHILDREN. Jill and Eustace learn more about eluding would-be captors than about French, math, or Latin; Jill has not learned COURTESY; and the children are unable to identify themselves as SON OF ADAM and Daughter of Eve, because they have never heard the STORY of Adam and Eve. What's more, they have never heard of the custom of swearing on a Bible to show good FAITH and have not been taught to use Christian names. Jill observes in Eustace's desire to explore Bism with RILIAN that he is more a product of Narnian ADVENTURES than a student of Experiment House. (SC 1, 3, 7, 13, 14, and 16)

F

FAIRY — A humanlike being with MAGICAL powers. Although fairies are popularly portrayed in literature as tiny winged creatures, before the sixteenth century they were thought to be the same size as human beings and thus much more dangerous. This is what Lewis had in mind when he writes that ANDREW Ketterley's godmother, Mrs. LEFAY, was one of the last mortals in ENGLAND to have fairy blood in her. DIGORY's only thought is that Mrs. Lefay was probably a bad fairy. (MN 2)

FAITH — Although C. S. Lewis uses the word "faith" only once in the *Chronicles*—in REEPICHEEP's challenge to CASPIAN to keep faith with all Narnians as their KING (VDT 16)—faith is a key concept in all of the chronicles.[1] Faith, as Lewis sees it, helps to keep reason alive and fighting when FEELINGS, circumstances, and IMAGINATION tempt one to abandon what one knows to be true, even about human beings, let alone about ASLAN. It is "the art of holding on to things your reason has once accepted, in spite of your changing moods."[2] There are a number of PROFESSIONS OF FAITH in the *Chronicles.*

[1]:Lewis uses "belief" six times and "believe" ninety-five times, about seventy percent of the time to express confidence by a human being or TALKING BEAST in another human being or Talking Beast (either their person or what they have said or done); the other uses denote confidence by a human being or Talking Beast in Aslan or the EMPEROR-BEYOND-THE-SEA.

[2]C. S. Lewis, *Mere Christianity,* Book III, Chapter 11, ¶5.

FARSIGHT — The eagle who brings TIRIAN the news of CAIR PARAVEL's capture and ROONWIT'S DEATH. His name reflects both the keen eyes of the bird of prey and his canny insight into character. Farsight is the biggest factor for victory in the second phase of the LAST BATTLE. He is among those who enter into NARNIA at the JUDGMENT, and he flies above the SEVEN FRIENDS OF NARNIA announcing that they are in the real Narnia. (LB 8, 9, 11, and 15)

FATHER CHRISTMAS — A huge, bearded man in a bright red ROBE whose appearance signals the end of the Hundred Years of Winter, during which time "it was always winter but Christmas never came." He is "big and glad and real," not just funny and jolly like the Father Christmas or Santa Claus we know in the modern world. He brings tools, not toys, to the CHILDREN and to Mr. and Mrs. BEAVER. (LWW 10)

[ASLAN'S BREATH; BIBLICAL ALLUSION(S).]

FATHER TIME — An enormous man with a waist-length white beard and a noble face, first encountered asleep in a cave in UNDERLAND (SC 10 and 14). The WARDEN OF THE MARCHES OF UNDERLAND tells PUDDLEGLUM that the old man, now Father Time, was once a KING in the

Upperworld and will wake only at the end of the world. He is indeed wakened at that time by ASLAN's roar and stands up in ETTINSMOOR. He blows a huge HORN, which causes the STARS to fall from the skies, signaling the END OF NARNIA. Nothing more is said of him. (LB 13 and 14)

[ASLAN'S VOICE; ESCHATOLOGY; SLEEP.]

FAUN(S) — In Greek and Roman MYTHOLOGY, small half-human woodland spirits, followers of BACCHUS and Pan. Linked with SATYRS, they are one of the NINE CLASSES OF NARNIAN CREATURES. They play wild, dreamy music on reed pipes and are described as having curly hair, little horns, and the legs and feet of goats. They are about the same size as DWARFS, but more slender and graceful. Their faces are described as mournful and merry at the same time, which is suggestive of the innate gravity of the GODS. Like the other mythological creatures (Dwarfs, NYMPHS, and the like), Fauns are not special creations of ASLAN but come into existence at the Lion's command for Narnia to awake.

The Fauns sent by PATTERTWIG to DANCE around CASPIAN (PC 6) are named Mentius, Obentinus, Dumnus, Voluns, Voltinus, Girbius, Nimienus, Nausus, and Oscuns. Their vaguely Latin names evoke qualities both of Fauns and pagan Roman culture. "Dumnus" is probably the diminutive of the Latin *dumus,* meaning "thicket, thornbush, bramble," in which Fauns might be expected to live. "Mentius" is from *mentior, mentitus,* meaning "deceiver, liar, cheater." Fauns are certainly clever seducers. "Nausus" is perhaps from *nasus,* meaning "nose," or from *nausea,* meaning "seasickness, indigestion." Fauns would be both inquisitive (nosy) and inclined to eat too much. "Nimienus," from *nimietas,* meaning "stuffed," is a reflection of Fauns' tendency to overindulge. "Oscuns" is perhaps from *osculare,* meaning "to kiss," or *oscitare,* meaning "to be half-asleep, to yawn." Fauns would sneak kisses and might also doze off after a feast. "Voltinus," perhaps from *volitare,* meaning "to fly about, flutter," suggests that Fauns are unstable pleasure seekers. "Voluns" is possibly from *volens,* meaning "willing, consenting," which suggests that Fauns were predisposed to give in to whatever their love of pleasure brings them. There is no known meaning for the names "Girbius" and "Obentinus." Three other Fauns are named in the *Chronicles:* Urnus, TRUMPKIN's ear trumpet–holding attendant, and JILL's attendant, Orruns. The derivation of their names is not known. The best-known Faun in the *Chronicles* is, of course, Mr. TUMNUS.

FEAR — In the *Chronicles,* Lewis teaches CHILDREN about the nature and the different kinds of fear. NIKABRIK says to CASPIAN, Doctor CORNELIUS, and TRUFFLEHUNTER, "Don't take fright at a name as if you were children" (PC 12). The name, however, is that of the WHITE WITCH; Lewis implies that healthy fear is good, and that we can trust a child's reaction to such a name as a STOCK RESPONSE to RIGHT AND WRONG. But according to ASLAN, SUSAN has listened to her fears and therefore her FAITH cannot come to the fore. Fear can also be determined by our preconceptions. The child at the NURSE's cottage in PC is not afraid of Aslan because he had never even seen a picture of an ordinary lion, but the TELMARINES are thoroughly terrified at the sight of Aslan because although they knew about lions, "they had not believed in [them] and this made their fear greater" (PC 15). Lewis teaches children about how to get a job done despite fear when he describes LUCY's walk through the HOUSE OF THE MAGICIAN. She has to walk by every room to get to the right room, and in every room an invisible or dead MAGICIAN may be hidden. "But," Lewis advises, "it wouldn't do to think about that" (VDT 10). Later, at the evening meal, Lucy notices how the fact that she is no longer afraid changes the look of everything upstairs that had frightened her earlier: "The mysterious SIGNS on the door were still mysterious but now looked as if they had kind and cheerful meanings" (11).

Similarly, SHASTA finds the tombs frightening and sinister during the night, but relatively harmless in the light of day (HHB 6). Such night fears are common to all people, and the crew of the *DAWN TREADER* instinctively does not want to venture to the DARK ISLAND (VDT 12). It is only COURAGE and the spirit of ADVENTURE that allow them to survive the ordeal of nightmare that awaits them.[1] For REEPICHEEP, yielding to fear is the greatest danger, and Lord RHOOP's condition shows what harm the extremes of fear can do: His is a voice "of one in such an extremity of terror that he had almost lost his humanity."

In SC, JILL and EUSTACE experience at least two phobias: Eustace's fear of heights (1) and Jill's fear of "twisty passages and dark places underground" (7), which intensifies in UNDERLAND. Indeed, much of SC deals with facing fears, and part of Jill's problem in continually forgetting the signs is an indulgence of her fear of taking responsibility. After the slaying of the QUEEN OF UNDERLAND and the fulfillment of Jill's original QUEST to rescue RILIAN, the prince

After their trek across the Great Desert, Aravis, Bree, Hwin, and Shasta plunge into a broad pool to cool off. (HHB 9)

calls on his companions to bid farewell to HOPES as well as fears (13). Healthy fear, says Lewis, and especially fear of the Lord, can help push us beyond what we think of as our limits. It is fear of a lion—actually, the Lion—that propels BREE and HWIN to gallop flat out (HHB 10).

In ASLAN'S COUNTRY, there is no fear. Aslan says to the newly created TALKING BEASTS, "Laugh and fear not" (MN 10). Aslan tells Lucy not to fear the river of DEATH, which lies across the way to his country, because he is "the Great Bridge Builder" (VDT 16). And in LB 16, in Aslan's country at last, Lucy observes that it is impossible to feel afraid, even if one tries to. Jill, free from fear at last, goes up the GREAT WATERFALL so high that if she *could* be afraid of heights she would be terrified, but here she can't be, and so the experience is thrilling.

[AUTOBIOGRAPHICAL ALLUSION(S); FEELING(S); IMAGINATION; NUMINOUS.]

[1]For further discussion of the Dark Island, see ADVENTURE and DREAM(S).

FEAR OF THE LORD — See NUMINOUS.

FEELING(S) — In a letter, C. S. Lewis admits: "My *mind* doesn't waver on that point: my feelings sometimes do."[1] In the *Chronicles*, Lewis teaches CHILDREN that feelings and emotions—even FEARS—are natural, acceptable, indeed, unavoidable (kissing and CRYING are not frowned upon in NARNIA), but that the children must not let feelings get them off their path. In LWW, PETER does not feel very brave—in fact, he feels nauseated. But that makes no difference to what he has to do. The theme of expression and acknowledgment of feelings runs through all the books. In PC, the children wish they could go on *feeling* as not-hungry as they did when they were still thirsty, and there is expression of feelings and emotions in their leave-taking of Narnia. In VDT, when CASPIAN shifts onto LUCY the responsibility of sailing into the darkness, she feels deeply that she would rather not but says otherwise. In HHB, HWIN shyly suggests that even though she and BREE feel that they can't go on, she knows that humans can spur HORSES on in spite of their feelings. And in LB, it is only when JILL and EUSTACE allow thoughts of the now-chaotic Narnia to come back into their minds that they feel desolate.

The characters in the *Chronicles* express a full range of emotions. Sadness is often expressed as "TERM-TIME" feelings (PC I); Caspian misses his NURSE intensely and cries a great deal; as the four children and TRUMPKIN grow more and more tired from rowing, their spirits fall; on DRAGON ISLAND Eustace feels sad and lonely (VDT 6) and the parting of the children from the *DAWN TREADER* is grievous; Jill feels sorrow at the eating of the Talking Stag and notices profound sadness on the faces of the EARTHMEN; DIGORY is miserable at the thought that his mother might die. The entire mood of LB is one of gloom, and at the news that TASH and ASLAN are one the ANIMALS feel depressed. But feelings of sadness are more than balanced by feelings of happiness, JOY, and ecstasy. In PC, feelings of terror and delight (the NUMINOUS) simultaneously burst into Caspian's mind, and after their meal, the children feel quite hopeful about the outcome of Caspian's campaign. In VDT 7, the undragonized Eustace is glad about the process of removal of his DRAGON skin—he loves his own body, puny though it may be. After drinking the sweet water of the LAST SEA, the drinkers feel "almost too well and too strong to bear it" (15). In SC 16, Jill cries out of sheer ecstasy at Aslan's beauty. In HHB 11, as SHASTA rides toward ANVARD in King LUNE's party, he feels happy for the first time since the day he entered TASHBAAN. The paradoxical joy of loneliness is mentioned several times in the *Chronicles:* Lucy regrets the *Dawn Treader* will not land at Felimath, because she likes to walk there. "It was so lonely—a nice kind of loneliness" (VDT 3). At the SILVER SEA, "even . . . the loneliness itself [was] too exciting" (VDT 16). In SC 6, Jill enjoys the loneliness of the moors. And in HHB 10, the HERMIT's enclosure is "a very peaceful place, lonely and quiet." There is also a place in Narnia for solemnity. Lucy has a solemn feeling as she takes her gift from FATHER CHRISTMAS (LWW 10); the Dawn Treaders experience the solemn feeling that they have sailed beyond the world (VDT 15). To Shasta, the HORNS at the opening of Tashbaan are "strong and solemn" (HHB 4); similarly, the preparations for battle with RABADASH are "very solemn and very dreadful" (13). Anger and other strong feelings are also expressed. Caspian dislikes his aunt, PRUNAPRISMIA (PC 5); DRINIAN's shortness with the KING is explained by the previous day's anxieties and the rainy weather makes Eustace disagreeable (VDT 8); Drinian's rage at REEPICHEEP is compared to that of a mother who is angry at her child for running out into traffic (VDT 15); at EXPERIMENT HOUSE Jill, interrupted in the middle of a good cry, flies into a temper (SC 1); Digory expresses righteous anger at his Uncle ANDREW's complete lack of principle and HONOR (MN 2); and POLLY is angry at the way Digory manhandles her in the great hall at CHARN (4). Lewis feels that anger can be taken too far, and his comment that talking horses, when angry, become more horsey in accent implies that anger dehumanizes us (HHB 1). In LB 2, TIRIAN unconsciously goes for his sword at ROONWIT's talk of lies, and at the whipping of the talking horse he is overcome by his feelings of grief and anger and rashly commits murder. Lewis comments that he and JEWEL are "too angry to think clearly."

[1]Letters II, 2 July 1949.

FENRIS ULF — A huge gray wolf, captain of the WHITE WITCH's secret police. His name is taken from the great wolf of Scandinavian MYTHOLOGY who was spawned by Loki—GOD of strife and spirit of EVIL—and who was slain by the fearless Vidar, son of Odin. In the British editions and the new HarperCollins editions of the *Chronicles,* the wolf is given the name "Maugrim," meaning perhaps "savage jaws," or alluding to *maugre,* "ill will."[1]

[1]No evidence has yet come to light to indicate whether or when Lewis made this change for his American readers. Since he liked Norse images, he may have thought the name change to be an improvement. Professor Thomas W. Craik disagrees: "The virtue of Maugrim . . . is that [it is] suggestive and not precise. Maugrim . . . has an unpleasant sound; 'grim' carries its sinister adjectival sense; 'mau' *might* be associated with 'maw,' but I think it more likely to connect with all Spenser's evil names beginning with Mal" (Letter to author dated July 31, 1979). See also WORLD ASH TREE.

FIRE-BERRY — A berry with the appearance of a little fruit or a live coal; it is too bright to look at directly (VDT 14). Every day a bird of morning carries one to RAMANDU and lays it in his mouth to rejuvenate him. The berries are said to grow in the valleys of the SUN, and it is conceivable that they grow from the same bush that produces the FIRE-FLOWERS from which LUCY's CORDIAL is made (although these are said to grow in the mountains of the sun).

FIRE-FLOWERS — The source of the juice used to make LUCY's healing CORDIAL (LWW 10). They grow in the mountains of the SUN and may be from the same plant that produces the FIRE-BERRIES that the birds of morning bring to RAMANDU to renew his youth (although these are said to grow in the valleys of the sun).

FIVE BLACK DWARFS — Suspicious, rather aggressive DWARFS who live in a cave near the SEVEN BROTHERS OF THE SHUDDERING WOOD. Unlike these Red Dwarfs, however, the Black Dwarfs do not seem to be brothers. They define their allegiance to CASPIAN negatively—they will support him if he is against MIRAZ, not because he is for Old NARNIA. That they are likely more sinister than the Red Dwarfs is evidenced by their wish to call for the help of OGRES and HAGS. (PC 6 and 7)

FLAMING MOUNTAIN OF LAGOUR — An awesome CALORMENE volcano. EMETH uses its force and fury as a simile to describe ASLAN's NUMINOUS quality (LB 15).

FLEDGE[1] — The winged HORSE in MN who carries DIGORY and POLLY to the GARDEN in the west; originally a cab-horse named Strawberry.[2] As Strawberry he is tired,[3] and when first encountered he is being VIOLENTLY flogged and maddened by Jadis. But his father was a cavalry horse, and in NARNIA his noble heritage is apparent. At ASLAN'S creation song, he looks and feels stronger, younger, and renewed. At Aslan's command, he becomes a winged horse named Fledge, the father of all flying horses. He is reunited with Digory and Polly at the GREAT REUNION in ASLAN'S COUNTRY.

[1]"To fledge" means "to gain the feathers necessary for flying" (reference provided by Dr. David C. Downing).

[2]In our world, horses named Strawberry are usually strawberry roan in color, a mixture of light brown and white hairs that gives the horse a pinkish tone.

[3]"Cab-horse tired" was an expression Lewis used to describe his dead-tired feeling, like the fatigue felt by the horses that pulled taxicabs and other delivery wagons before the automobile.

"FOR THE FIRST TIME" — These words signal that an enormous change is taking place in a character. In LWW 11, EDMUND feels sorry for the feasting ANIMALS turned into stone, and "for the first time in this STORY felt sorry for someone besides himself." In VDT 5, EUSTACE, upon leaving the group and wandering around DRAGON ISLAND alone, "began, almost for the first time in his life, to feel lonely." He later acts bravely "for the first time in his life" when he attacks the SEA SERPENT (8). On RAMANDU'S ISLAND, when the crew first looks at the DAUGHTER OF RAMANDU, they think that they have never before known what beauty means (13). In SC 13, Eustace and JILL use each other's Christian names "for the first time" in their departure from the witch's castle. In HHB 7, ARAVIS, tired of LASARALEEN's silliness, "for the first time ... began to think that traveling with SHASTA was rather more fun than . . . life in TASHBAAN." When Shasta slides off BREE to help Aravis, "he had never done anything like this in his life before." (10)

FRANK I — The first KING of NARNIA; he is a former London cabdriver and the husband of HELEN. The humble, good-hearted Frank is a good and kind ADULT, and his coronation is the fulfillment of the prophecy that "the last shall be first."[1] King Frank is last seen in LB, sitting with his wife on their thrones in the GARDEN of the west. Lewis's comment that TIRIAN feels, in their presence, as we would feel if we were in the presence of Adam and Eve,[2] is a strong suggestion that this is their role in Narnia.

[1]Mark 10:31.

[2]Cf. Ransom before Tor and Tinidril in Lewis's *Perelandra*, Chapter 17.

FRIEND(S), FRIENDSHIP — C. S. Lewis believed that friendship is one of the four major kinds of love that human beings can experience.[1] Friendship usually arises from companionship, spending time together. When two companions discover that they think and feel almost the same way about the same thing, they discover that they are friends. Unlike family love and sexual love, friendship is almost never jealous: Healthy friendship welcomes the addition of a third person, a fourth person (and so on)—as long as the new friends are just as passionate about the "thing" the original friendship is about. The dangers that friendship is exposed to, and for which therefore it needs the gardening of divine love, are exclusivism and friendship in an EVIL attitude or enterprise.

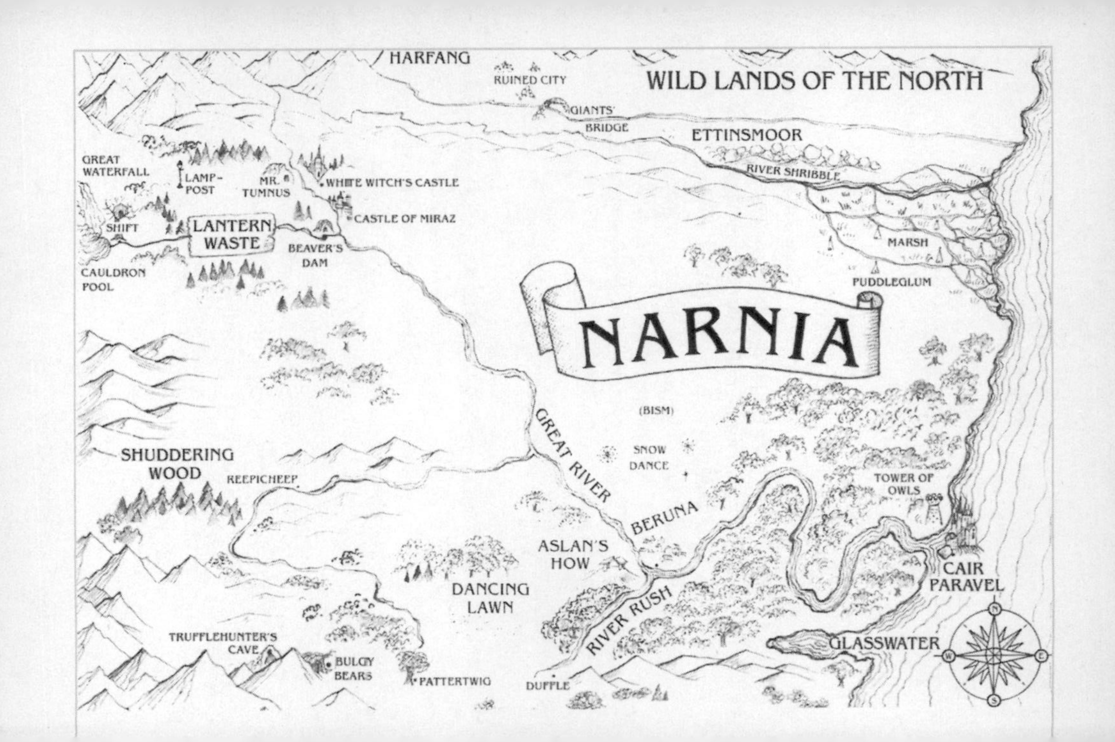
NARNIA
WILD LANDS OF THE NORTH
HARFANG
RUINED CITY
GIANTS' BRIDGE
ETTINSMOOR
RIVER SHRIBBLE
MARSH
PUDDLEGLUM
GREAT WATERFALL
LAMP-POST
MR. TUMNUS
WHITE WITCH'S CASTLE
CASTLE OF MIRAZ
SHIFT
LANTERN WASTE
BEAVER'S DAM
CAULDRON POOL
(BISM)
GREAT RIVER
SNOW DANCE
TOWER OF OWLS
SHUDDERING WOOD
REEPICHEEP
BERUNA
ASLAN'S HOW
CAIR PARAVEL
DANCING LAWN
RIVER RUSH
GLASSWATER
TRUFFLEHUNTER'S CAVE
BULGY BEARS
PATTERTWIG
DUFFLE
N
W
E
S

MT.
PIRE
ANVARD
PASS
STORMNESS
ARCHENLAND
HERMIT'S
DWELLING
RIVER WINDING ARROW
GORGE
GREAT DESERT
ANCIENT
TOMBS
TASHBAAN
CALORMEN
COTTAGE
OF ARSHEESH

THE VOYAGE OF THE DAWN TREADER
EASTERN OCEAN
SILVER SEA
CAIR PARAVEL
GALMA
TEREBINTHIA
MUIL
BRENN
SEVEN ISLES
LONE ISLANDS
AVRA
DOORN
FELIMATH
DRAGON ISLAND
BURNT ISLAND
DEATHWATER
LAND OF THE DUFFERS
DARK ISLAND
RAMANDU'S ISLAND
SEA PEOPLE
N
E
S
W

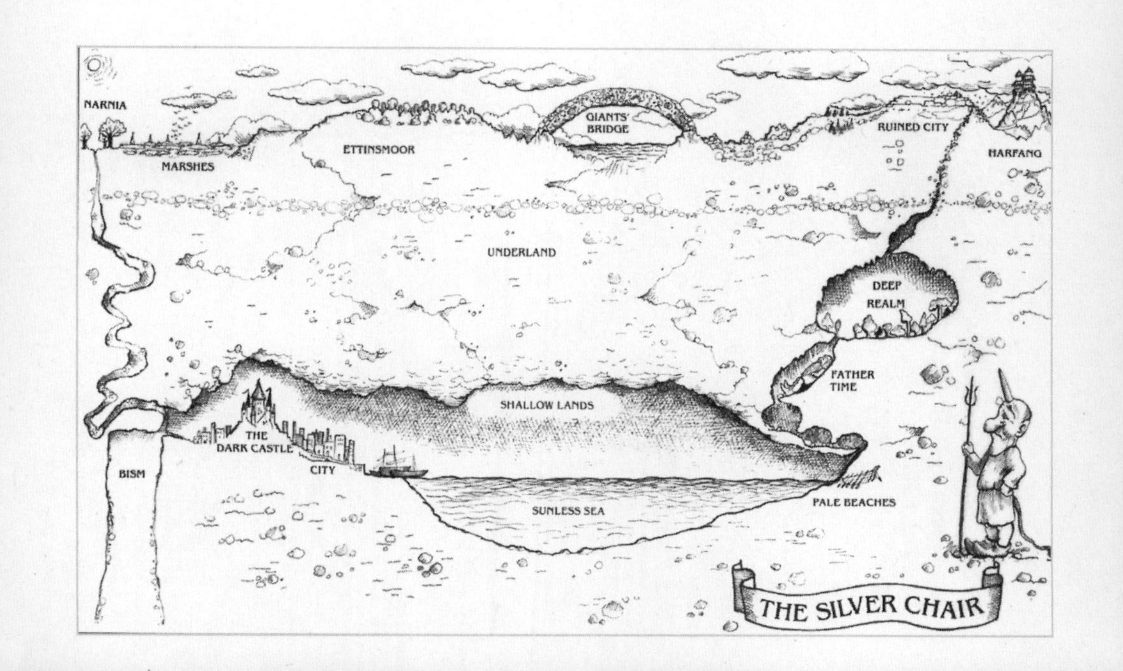
NARNIA
MARSHES
ETTINSMOOR
GIANTS' BRIDGE
RUINED CITY
HARFANG
UNDERLAND
DEEP REALM
FATHER TIME
SHALLOW LANDS
THE DARK CASTLE
CITY
BISM
SUNLESS SEA
PALE BEACHES
THE SILVER CHAIR

LUCY and Mr. TUMNUS, REEPICHEEP and Doctor CORNELIUS, Reepicheep and the DRAGON EUSTACE, SHASTA and ARAVIS, Prince RILIAN and his father's friend Lord DRINIAN. POLLY and Queen HELEN, POLLY and DIGORY, King TIRIAN and JEWEL the UNICORN—these are just a few of the interspecies, intergenerational friendships in the *Chronicles*. Lucy also has a rarer experience of instant friendship (arising from no previous companionship): She spontaneously loves one of the SEA PEOPLE, the fish-herdess.

Lewis himself had some excellent friends (see AUTOBIOGRAPHICAL ALLUSION[S]).

[1]The other two natural kinds of love are affection (family love) and sexual love; these three loves are the garden of human love and they need divine, selfless love to tend this garden. See Lewis's book *The Four Loves.*

G

GALE — Ninth KING of NARNIA, a direct descendant of King FRANK and Queen HELEN (LB 8). He kills the DRAGON that has long terrorized the inhabitants of the Lone Islands. In their gratitude they hail him as their emperor, a title he passes on to all subsequent kings of Narnia. His whole STORY is one of the UNFINISHED TALES of Narnia.

GALMA, ISLE OF — Located a day's voyage north of CAIR PARAVEL. Galmians are noted for their nautical expertise. Since MIRAZ had forbidden his people the study of navigation, Galma provides a ship and crew for the SEVEN

NOBLE LORDS sent by the usurper to explore the ocean east of NARNIA. Pug also forces Galmians to man his SLAVE ships. (PC 8 and VDT 2, 3, and 14)

GANG, THE — The ten or fifteen bullies of EXPERIMENT HOUSE who terrorize JILL, EUSTACE, SPIVVINS, and other well-behaved CHILDREN in SC. A partial list includes Adela Pennyfather, Cholmondele[1] Major, Edith Winterblott,[2] "Spotty"[3] Sorner, Bannister, the Garrett twins, and CARTER. Edith Jackle is "not one of them herself but one of their hangers-on and tale bearers." These despicable children are special friends of the HEAD, who sees them as interesting psychological cases and never punishes them. They intimidate the others with threats of PAIN and torture (hinted at by Carter's cruelty to the rabbit). At the end of SC 16, they get their comeuppance in a sequence that must be the DREAM of every child who has ever been picked on by a bully.

[1]For American readers who may not know, this is pronounced "Chumley."

[2]Note the delightfully suggestive quality of her name. SCHOOL and TERM-TIME, already blots on the year, are made even more horrible by people such as Edith Winterblott.

[3]He may have been called "Spotty" because he had freckles; it is also likely that he had acne, because in England pimples are called "spots."

GARDEN — The spiritual heart of ASLAN'S COUNTRY; it may be reached by going to the Utter East or by flying over the WESTERN WILD. Because the descriptive details do not harmonize exactly, it is not clear whether Lewis intended the garden in VDT 7 to be the same as the garden in MN 12 and 13 and LB 16. EUSTACE is de-dragoned by ASLAN in a mountaintop garden and renewed in a bathlike well with marble steps going down into it. The garden in MN is located at the top of a steep, green hill, and it has a wonderful SMELL. Inside a high wall of green turf that surrounds the garden grow TREES with green, bluish, and silvery leaves. Its high gates of gold face the rising SUN. It is here that DIGORY is sent to pick the silver apple from which will grow the TREE OF PROTECTION. The good MAGIC of this place is in strong contrast with the bad magic of the GOLDEN BELL AND HAMMER. The garden has an absolute feeling of PRIVACY and is filled with a sacred SILENCE, except for the water gently splashing in a fountain near its center. It is far larger inside than it seems outside, a quality that is also found to be true of Aslan's country in LB.

[GEOGRAPHY, NARNIAN.]

GARRETT TWINS — See GANG, THE.

GEOGRAPHY, NARNIAN — NARNIA is the overall name for a flat world that comprises three countries: ARCHENLAND and Narnia and the wild wastelands to the north of the great desert, and CALORMEN to the south. All are bordered on the east by the Eastern Sea and on the west by the WESTERN WILD, beyond which is the GARDEN. Geographically, Narnia is reminiscent of our own world: Calormen is akin to the Near and Middle East, Narnia itself is much like Great Britain and Scandinavia, and Archenland seems topographically like Switzerland. Since general directions in Narnia seem to be flipped east-for-west from those in Europe, Narnia (on the eastern shore) and CAIR

PARAVEL (on the eastern tip of Narnia) are a fair match for Britain (to the west of Europe) and Wales (with its Arthurian sites—at the western tip of Britain).[1]

Traveling east, one gets closer and closer to ASLAN'S COUNTRY; indeed, the east has a very mystical landscape. The breakers are long and slow, a retired STAR and his DAUGHTER live on an enchanted island, the sea is filled with lilies, the water is sweet, and the rising SUN is large and close. And, because Narnia is flat, those who continue to the Utter East disappear over the edge, presumably into ASLAN'S country. To the west,[2] however, is another entrance into Aslan's country; it is here that the garden houses the TREE OF PROTECTION. This seeming contradiction is resolved at the end of LB, when, standing in Aslan's country, LUCY, PETER, and EDMUND can see that Narnia is ringed by a range of mountains, called the mountains of the Utter East on one side and the Western Wild on the other side. And the garden in which they stand is really in the middle of the ring: It is the spiritual heart of Aslan's country.

[1]The insights of the last sentence in this paragraph are those of John Singleton.

[2]Probably because the sun sets in the west, "going west" is a euphemism for dying; thus "he has gone west" means "he died." See John Ciardi, *A Browser's Dictionary* (New York: Harper & Row Publishers, 1980), 410.

GHOST(S) — The spirits of dead persons, which remain to haunt the living. (LWW 15, PC 3, HHB 11, and SC 16)

GHOUL(S) — Evil spirits, often associated with graves and corpses, invited by the WHITE WITCH to attend the slaying of ASLAN (LWW 13). The word *ghoul* is Arabic in origin, a fact very appropriate in light of the relatively frequent mention of the FEAR of ghouls that SHASTA, Aravis, and LASARALEEN's groom experience in the shadows of the ancient tombs (HHB 3, 6, and 9). This is another indication that Lewis's fictional CALORMEN is drawn largely from Britain's experience of its Middle Eastern colonies.

[RACISM AND ETHNOCENTRISM.]

GIANT(S) — A race of tall beings that appears in virtually all world MYTHOLOGIES. Giants are one of the NINE CLASSES OF NARNIAN CREATURES as related by Doctor CORNELIUS. In NARNIA, there are roughly two types of giants: good and bad. Both types use clubs as weapons and wear knee-high spiked boots. The bad giants live generally in the north and are hostile and even murderous toward humans, DWARFS, and TALKING BEASTS. They include the so-called Gentle Giants of HARFANG; the fierce giants of northern Narnia who are driven back by the KINGS and queens in LWW; the Northern giants who were defeated by CASPIAN X in the summer of 2304 N.Y.; and the ETTINS, who are part of the WHITE WITCH's forces against ASLAN. Good giants include RUMBLEBUFFIN (and the entire Buffin clan), WIMBLEWEATHER, and Stonefoot. Good or bad, giants are ugly and not very intelligent. Some giants were apparently once capable of greater achievements, as indicated by the ruined city of the ancient giants, with its well-engineered bridge and the frieze work that decorates the balustrade.

GIFT OF SPEECH — When ASLAN changes some of the ANIMALS into TALKING BEASTS, he gives them the land of NARNIA and responsibility for the DUMB BEASTS from which they were taken. Though this would seem to put the Talking Beasts on an equal footing with the humans, they acknowledge the SONS OF ADAM and Daughters of Eve as their superiors.[1]

[1]See HIERARCHY.

GIFT(S) — See FATHER CHRISTMAS.

GINGER[1] — An orange-colored tomcat who has FAITH neither in ASLAN nor in TASH and volunteers to enter the STABLE but soon runs out, completely disheveled and shrieking in terror. He has seen the real Tash, and—for his heresy—has been deprived of the GIFT OF SPEECH by Aslan. He grows less and less like an intelligent TALKING BEAST, and more and more like a wild DUMB BEAST and is never seen again. (LB 3, 7, and 10)

[1]*Letters to an American Lady,* 22 February 1958, mentions that one of Lewis's cats was "a huge Tom called Ginger" (reference supplied by David C. Downing).

GLASSWATER CREEK, CREEK(S) — The term *creek* as used by Lewis is often quite confusing to American readers. In England, a creek is a saltwater inlet or bay, not a small freshwater river. So Glasswater Creek is a small inlet or bay south of CAIR PARAVEL, the head of which is just behind the Hill of the STONE TABLE (PC 8 and 9). SHASTA and ARSHEESH also live on a "creek" of the sea (HHB 1), and BREE and Shasta and the unknown HORSE and rider are forced out into a creek (HHB 2), where they begin to talk.

GLENSTORM — A CENTAUR in PC (6, 13, 14, and 15). The father of three sons, he is a noble creature with glossy chestnut flanks and a full golden beard. Being a prophet and stargazer, he knows in advance the mission of CASPIAN, TRUFFLEHUNTER, TRUMPKIN, and NIKABRIK. He accompanies EDMUND and WIMBLEWEATHER when they deliver PETER's challenge to MIRAZ and is appointed one of the marshalls of the lists. Glenstorm's life role is "to watch," just as Trufflehunter's (and all badgers') is "to remember." He is among those summoned to the GREAT REUNION in LB. His name suggests that he is a force to reckon with, whose sheer weight makes its presence felt on the pastures.

GLIMFEATHER — A huge white owl in SC (3, 4, and 16), he is about as tall as a good-sized DWARF (about four feet tall). His name is derived from the Middle High German word *glim,* which means "to shine or gleam." Thus he is a bird whose white feathers glow, especially in the moonlight under which he so often flies. He is present for the GREAT REUNION in LB.

GLOZELLE — A scheming TELMARINE lord, friend of SOPESPIAN, and counselor to MIRAZ and his marshall of the lists (PC 13 and 14). With characteristic Telmarine treachery, he stabs Miraz to death because of a personal insult.

GNOMES — See EARTHMAN, EARTHMEN.

GOD(S) — Specified by Doctor CORNELIUS as one of the NINE CLASSES OF NARNIAN CREATURES, these MYTHOLOGI-

CAL beings are present throughout the *Chronicles.* In fact, the daughters of King FRANK and Queen HELEN marry wood gods and river gods. Narnian gods, goddesses, and semi-deities include BACCHUS, FAUNS, DRYADS, HAMADRYADS, MAENADS, NAIADS, NYMPHS, POMONA, the RIVER-GOD, SATYRS, SILENUS, SILVANS, TREE-PEOPLE, and WOOD PEOPLE. CALORMENE gods and goddesses include TASH, Azaroth, and ZARDEENAH. The question arises, of course, as to what pagan gods and goddesses are doing in a Christian universe. Lewis replies that it is only in God's name that the spirits of nature can rule their domains with "beauty and security." Without God, they would disappear or "become demons."[1]

[1]C. S. Lewis, *The Four Loves*, 166.

GOLDEN AGE OF NARNIA — The name given to the time of the reign of the KINGS and queens, from 1000 to 1015 N.Y.,[1] in which the ADVENTURES of SHASTA and ARAVIS take place. It is ended when EDMUND, LUCY, SUSAN, and PETER vanish from NARNIA during the hunt for the WHITE STAG. During the Golden Age, all that is best of Narnian life is in full bloom: There are many more TALKING BEASTS than ever afterward; Narnia is a great sea power; the smaller woodland people and ANIMALS of Narnia feel so safe and secure that they get a bit careless, even ignoring the CALORMENE threat; the fierce GIANTS are driven away; and Narnia forms peaceful, friendly alliances with countries "across the sea." (PC 4 and 5; VDT 2, 13, and 15; HHB 1 and 12; SC 3; and LB 4)

[1]See Lewis's outline of Narnian history in *The Land of Narnia*, 31.

GOLDEN BELL AND HAMMER — A golden bell and a little hammer enchanted by Jadis stand on a four-foot-high square pillar in the middle of the royal hall at CHARN (MN 4). The MAGIC in the hall enables DIGORY and POLLY to understand the meaning of the writing on the pillar,[1] which beckons them to lift the hammer and ring the bell, which Digory does. A SWEET note rings louder and louder until it culminates in an EARTHQUAKE that destroys the room.

[WHITE WITCH.]

[1]The verse on the pillar represents the seduction of EVIL by dissipation, whereas the verse on the GARDEN of the west, which Digory later faces, represents the attraction to good through OBEDIENCE.

GOLDEN TREE — A TREE made of real gold that grows from two half-sovereigns (about the size of a U.S. quarter) that fall out of Uncle ANDREW's pockets into the rich Narnian soil. Along with the SILVER TREE, it grows inside the cage in which Andrew is kept (MN 11 and 14).

GOLG — An EARTHMAN captured by PUDDLEGLUM after the DEATH of the QUEEN OF UNDERLAND (SC 13). He is three feet tall and has little pink eyes, a hard ridge on his head, and the look of a pigmy hippopotamus. When Golg first encounters RILIAN, EUSTACE, JILL, and PUDDLEGLUM, he thinks they are out to do him VIOLENCE in the name of the queen, but the four adventurers reassure him that they have killed her, and then he tells them about his people and their way of life. He joyfully tells his people that the queen is dead and that these travelers can be trusted. He invites the four to visit the land of Bism, where he promises to show them the fiery SALAMANDERS. Golg then shows them the way to the surface, wishing them well.

GREATNESS OF GOD, THE — As CHILDREN grow older, they discover that the house they grew up in and the familiar sights and people of their childhood have all grown smaller. In a remarkable reversal of this expectation, LUCY, when she meets with ASLAN in PC 10, comments to him that he seems to have grown bigger. The Lion tells her that he has not grown larger but that, because she has grown, she finds him bigger. He seems to state a principle that as persons continue to grow, they will continue to find him bigger. This revelation makes Lucy so happy that she is SILENT. Lewis's intention in this scene is twofold: A person's growth in spiritual maturity will always be matched by God's growth in meaning for that person, and yet we will never have a comprehensive knowledge of God.

Lewis comes at this same insight another way in the discussion between Shasta and the Large Voice in the night fog (HHB 11). SHASTA asks the Voice if it is a GIANT. It responds that in a certain sense it is a giant but not like the creatures Shasta has in his fearful IMAGINATION. But this part of the picture needs the complement of the other experiences of God that Shasta has that night: the God of PROVIDENCE, intimately involved in a person's HISTORY, gloriously strong and beautiful and wise.

Theologians use the term *transcendence* to describe the otherness, the "largeness," and the "overness" of God with respect to all creation. This term is often used as a complement to or in contrast to *immanence,* the "hereness" and "withinness" of God. This pair of terms must be kept in balance. If God's transcendence is stressed (as in some Eastern religions, in Islam, and in some Christian churches), then a way is open for a divorce to be decreed between God and what he has made, and creation to be devalued. If God's immanence is stressed (as in pantheism, where everything is divine; or as in naturalism, where nature is said to have, within her, her own explanation of her existence), then God is in danger of being relegated to the history of ideas.

[ASLAN'S BREATH; EMPEROR-BEYOND-THE-SEA; TRINITY.]

GREAT REUNION — The reunion of all the good characters in the *Chronicles* takes place in four stages in LB: TIRIAN is reunited with the SEVEN FRIENDS OF NARNIA, who include his helpers JILL and EUSTACE (12); these eight are joined by their comrades-at-arms from the LAST BATTLE (14); these meet the heroes from the WAR OF DELIVERANCE and from the GOLDEN AGE OF NARNIA (16); and all of these go on to meet ASLAN and Mr. and Mrs. PEVENSIE (16).

GREAT RIVER — The major river of NARNIA, which flows out of CALDRON POOL and runs from west to east across the valley. Created by ASLAN on the first day, it is ruled by the RIVER-GOD and inhabited by his daughters, the NAIADS, in response to the command "be divine waters."

GREAT WATERFALL[1] — The farthest western limit of NARNIA and the source of the GREAT RIVER. It falls from enormously high cliffs and pours into CALDRON POOL. FARSIGHT leads the company to the waterfall in LB, and JEWEL is the first to swim *up* the waterfall to ASLAN'S

COUNTRY, followed by TIRIAN, EUSTACE, the DOGS, and JILL.

[1]Waterfalls are a recurrent image in Lewis's writing. For example, in *The Great Divorce,* Chapter VI, the lovingly described waterfall is also "a bright angel who stood, like one crucified, against the rocks and poured himself perpetually down towards the forest with loud joy."

GREAT WOODS — The Old Narnian name for the forest that ranges from the Eastern Sea inland as far as the Hill of the STONE TABLE, and from above CAIR PARAVEL on the north to the southern mountains that are the border between NARNIA and ARCHENLAND. During the reign of the TELMARINE conquerors, this forest is renamed the Black Woods when the invaders begin to FEAR the TREES. The awakened trees are the deciding factor in the final battle of the WAR OF DELIVERANCE, when they chase the army into the GREAT RIVER. As a result of this victory, the forest reclaims its original name. (PC 4, 7, and 14)

GREED — The impulse toward greed—the combination of the vices of pride, avarice, and gluttony—is almost always the downfall of the overtly wicked characters, but it is also a temptation faced (and overcome) by the good characters of the *Chronicles.* Uncle ANDREW becomes involved in dark MAGIC because of his greed for power, and he is unable to see NARNIA's glory and is greedy for what its fertile soil can bring him commercially. All of the witches—the WHITE WITCH, Jadis, and the QUEEN OF UNDERLAND—are greedy for power over other worlds, be they Narnia or our world. The CALORMENES are perennially greedy for power over Narnia. EDMUND's greed for more TURKISH

Delight causes him to betray his brother and sisters; Eustace's greed for the dragon's gold causes him actually to become a dragon. The flush of greed comes over Caspian's face on Deathwater Island as he dreams about the power this magic water might bring him (akin to King Midas's golden touch). Growing more greedy every moment, he claims the island for Narnia for all time and swears all to secrecy on pain of death. The argument grows greater until Aslan erases their memories; they remember only that something terrible is connected with the place, and they shun it.

GREENROOF — One of the summer months of the Narnian calendar (PC 13). The name suggests the lush foliage of the season.

GRIFFLE — A cynical Black Dwarf who is the spokesman for the majority of Dwarfs released by Tirian. He has no faith in Aslan and declares that the Dwarfs are independent of any allegiance. Griffle is present for the final assembly at Stable Hill. In the struggle between the Dwarfs and the Calormenes, Tirian hears Griffle using "dreadful language." Because Diggle is the spokesman for the Dwarfs who survive the last battle to be thrown into the stable, it is presumed that Griffle was killed in the battle. (LB 7, 10, and 12)

GUMPAS — Governor of the Lone Islands, called "His Sufficiency." Gumpas is the epitome of the petty civil servant, immersed in detail and unmindful of the larger issues. He is completely immersed in accounts, forms, rules, and regulations. He is removed from office by CASPIAN and replaced with Lord Bern (PC 3 and 4). The word *gumpus* means "a foolish person."

GWENDOLEN — A schoolgirl who disrupts her class by calling attention to the Lion outside (PC 14). ASLAN invites her to join in the REVELRY, calling her "sweetheart" (a possible allusion to Lewis's belief that God loves us first, even before we recognize him). She is overjoyed at Aslan's invitation and immediately joins the DANCE of the MAENADS. They help her take off her unnecessary and uncomfortable clothes, an action suggestive of Lewis's admitted abhorrence of SCHOOL clothes as the ultimate restraints.
[AUTOBIOGRAPHICAL ALLUSION(S); PRIZZLE, MISS; ROBE(S), ROYAL; UNIVERSALISM.]

H

HAG(S) — Ugly, old, EVIL witches summoned by the WHITE WITCH to the slaying of ASLAN at the STONE TABLE (LWW 13, 14, and 17). Four Hags bind the Lion, and stand one at each corner of the table, holding torches. Even after the crowning of the four KINGS and queens, rumors of Hags are heard. In PC 6 and 12, the FIVE BLACK DWARFS are in favor of asking the aid of a Hag up in the crags, and NIKABRIK's friend at the council of war is an obsequious Hag who claims only minor MAGICAL skills, which she will gladly use against MIRAZ. She whines that the White Lady (her name for the White Witch) is not dead—nobody's ever heard of a witch that died (see EVIL). She is beheaded by TRUMPKIN.

HAMADRYAD(S) — In Greek and Roman MYTHOLOGY, wood NYMPHS who live and die with the TREE of which they are the spirits. They are mentioned in conjunction with DRYADS in PC 9 and 14 by LUCY and TRUFFLEHUNTER. Their creation by ASLAN is not specified; they appear to have come into being in response to Aslan's command for NARNIA to awake, and for the trees to be waking trees. They are therefore part of the wild people of the wood, the TREE-PEOPLE and the WOOD PEOPLE. No specific individual Hamadryad or its activity is mentioned in the *Chronicles.*

HAMMER, THE — See LEOPARD, THE.

HARDBITERS — The family name of three badgers summoned to and present at CASPIAN's great council, which plans the WAR OF DELIVERANCE (PC 6 and 7). The name "Hardbiter" evokes the fierce tenacity that is an important ingredient in the hieroglyph "badger."[1]

[1]See TALKING BEAST(S).

HARFANG — The name of a mountain and of a castle inhabited by EVIL* GIANTS where, the Lady of the Green Kirtle says, PUDDLEGLUM, EUSTACE, and JILL will find COMFORT (SC 6–9). "Harfang" ("rabbit catcher") is the name of the great snowy owl.

HAROLD SCRUBB — The husband of ALBERTA Scrubb and father of EUSTACE (VDT 1, 2, and 5). Eustace addresses him as Harold, a custom contrary to the attitude of COURTESY that Lewis felt CHILDREN should have toward ADULTS, especially parents.

HASTILUDE — A medieval form of spear-play; a kind of tournament that was almost as dangerous as war itself. RABADASH excels at this form of entertainment (HHB 4).

HEAD — The unnamed woman who is the SCHOOL principal of EXPERIMENT HOUSE. She favors the bullies in the GANG, finds them interesting psychological cases, and makes them her special friends. (SC 1 and 16)

HELEN — First queen of NARNIA, wife of KING* FRANK I; before coming to Narnia she was a London housewife named Nellie. She is transported into Narnia by ASLAN, at Frank's wish that his wife be there too. She gives birth to at least two boys and two girls, and her second son becomes the first king of ARCHENLAND.[1] She is last seen on her throne with King Frank in the GARDEN of the west at the GREAT REUNION.[2] (MN 11, 12, 14, and 15)

[1]For a different story, see COL.

[2]See FRANK I, n. 2.

HELL — Although hell is not specifically mentioned in the *Chronicles,* it can be inferred in at least two instances. In LB 12 and 13, TASH is dismissed to "his own place," which is implicitly a place of permanent punishment. Since heaven, a place of infinite JOY, corresponds to ASLAN'S COUNTRY, it may be assumed that Tash's place corresponds to hell. DIGGLE the DWARF speaks for his fellows when he calls the inside of the STABLE a "pitch-black, pokey, smelly little hole," a hell of his own making.

HERMIT OF THE SOUTHERN MARCH — A 109-year-old man who lives on the Southern MARCH, or border, of ARCHENLAND, north of the great desert. The barefoot hermit is tall and bearded, dressed in an ankle-length ROBE, and he leans on a straight staff. His enclosure is perfectly circular (the very picture of the divine wholeness of the place) and his pool—in which he can see events happening elsewhere—is perfectly still (symbolic of the quality and depth of his PRAYER). A huge, beautiful TREE stands at one end, and the finest grass covers the ground. He lives in an old stone, thatch-roofed house, and keeps goats. In appearance, he is very like two other old men, RAMANDU and CORIAKIN. Curiously, although these two are retired STARS, the hermit is not; but his presence must be equally as commanding as theirs, for SHASTA mistakes him for the KING of Archenland. He is a MAGICIAN of sorts and by his art can read the present but not the future. He loves ANIMALS, calling the goats his cousins and giving BREE and HWIN rubdowns worthy of a king's groom. During a conversation with ARAVIS about her strangely superficial wounds (which she attributes to luck), the hermit says that he has never met the thing called luck.[1] He admits he doesn't understand the meaning of it all, but he exercises *studiositas*[2] and HOPE: "If ever we need to know the meaning, you may be sure that we shall." (HHB 10, 11, 13, and 14)

[1]See Providence.

[2]See Curiosity.

HIERARCHY — There is an ascending order of nobility among creatures. In the medieval vision everything is headed by God, the source of all being. From that unreachable height one crosses over into creation and the scale works itself down from the highest angel to the lowest inanimate material being. Christ has a special place in the whole system since he is on both sides of the great divide between Creator and creature.

In Narnia, the hierarchy is topped by the transcendent Emperor-beyond-the-Sea. Aslan—as Son of the Emperor-beyond-the-Sea, Lord of the Wood, and King of Beasts—has one foot in Narnia and one foot in the realm across the sea. Ultimately, he is the bridge between the two realms; this becomes clear at the end of VDT (16).[1] Aslan is followed in nobility by the human children, youth, and adults, who are in turn followed by the Talking Beasts and the dumb beasts.

The importance of rightful authority and its acknowledgment in honor, obedience, and courtesy—all of these elements within the *Chronicles* show us a world that is shot through with the principle of hierarchy. In LWW, the question of their human heritage is put to the children three times: first, in the meeting between Tumnus and Lucy (2); second, when Edmund encounters the White Witch (4); and finally, when Mr. Beaver establishes contact with the four children (5). When they are introduced to Mrs. Beaver, her awed response raises our expectations of these children precisely as humans, as descendants of

Adam and Eve.[2] We then learn, as Mr. Beaver quotes the old prophecies, that Aslan's return to "put all to rights" (8) will be accompanied by the appearance of four human children and their enthronement at CAIR PARAVEL.[3]

In PC 5, CASPIAN is acknowledged as the rightful ruler of Narnia by the Talking Beasts, not only because of his immediate lineage, but also because he is a human and a SON OF ADAM. "Narnia was never right except when a son of Adam was KING," TRUFFLEHUNTER says in response to NIKABRIK's hatred and suspicion of humans. This acknowledgment is seconded by the THREE BULGY BEARS and then by the great CENTAUR* GLENSTORM, who announces that "a Son of Adam has once more arisen to rule and name creatures" (6). In MN 11, we learn about King FRANK and Queen HELEN—the cabby and his wife who are chosen to be the first rulers of Narnia by Aslan. "You shall rule and name all these creatures, and do justice among them, and protect them from their enemies when enemies arise."

In all these instances it is the humans who are clearly born to rule and to make HISTORY; how they choose to respond in the face of Aslan's initiatives is decisive for the history of Narnia, though Aslan always has the last word. It is, for example, a human choice that brings EVIL into Narnia at its creation, setting the stage for both Edmund's betrayal and Aslan's suffering in LWW. In the final chapter of MN, Aslan warns the children of the destructive possibilities of human freedom. A whole world—CHARN—has been blotted out through the evil choices of Empress Jadis: "Let the race of Adam and Eve take warning."

The dwarf brothers Rogin, Bricklethumb, and Duffle serve the starving Shasta his first Narnian meal. (HHB 12)

[1]But the comparison to Christ must not be pushed too far here. Aslan comes on the scene always and already a lion: He didn't become incarnate as a lion in order to save Narnia.

[2]There even seems to be a faint echo of Simeon's *nunc dimittis* from the Gospel of Luke (2:29–32). "At last! To think that ever I should live to see this day!" says Mrs. Beaver. See BIBLICAL ALLUSION(S).

[3]The titles Son of Adam and Daughter of Eve are again invoked when FATHER CHRISTMAS fits Peter, Lucy, and Susan out with gifts appropriate to their respective tasks in the struggle against the evil witch.

HISTORY — Commonly experienced in SCHOOL as dull books about wars, battles, and dates. The schoolmistress of the unnamed town near Beruna is in the midst of teaching a history class when ASLAN comes by, and Lewis comments that history as taught under MIRAZ's rule was "duller than the truest history you have ever read, and less true than the most exciting ADVENTURE* STORY" (PC 14).

In contrast, Lewis says that what is really important is the peace between wars, and the lives of ordinary people. When DIGORY wishes to know the names of the places in the newly created NARNIA that he and POLLY are seeing from the air, Polly reminds him that they won't have names until there are people in those places. Digory is excited at that because then there will be histories, which Polly (speaking for Lewis) equates with the very dull learning of battles and dates (MN 12). In LB 8, JEWEL comments that history is made up of big events, but that the times of peace flow on with no chronology. He explains to JILL that there's *not* "always so much happening in Narnia," and that in between the visits of the English CHILDREN there were "hundreds and thousands of years . . . in which there was really nothing to put into the history books."

HOLIDAYS — See TERM-TIME.

HOLY SPIRIT — See ASLAN'S BREATH.

HONOR — To seek honor in OBEDIENCE and to behave honorably may be said to be one definition of a true Narnian. In PC 15, CASPIAN is ashamed that he comes of such a dishonorable (TELMARINE) lineage. ASLAN replies, "You come of the Lord Adam and Lady Eve. And that is both honor enough to erect the head of the poorest beggar and shame enough to bow the heads and shoulders of the greatest emperor on earth. Be content."

HOPE(S) — In his book *Mere Christianity,* C. S. Lewis defines uppercase-h Hope[1] as "a continual looking forward to the eternal world."[2] This simple definition links hope to two other deep themes in Lewis's life and thought: JOY and LONGING. The definition also distinguishes supernatural hope from lowercase-h hope—more often in the plural, "hopes"—the limited assurances and expedients that humans count on,[3] as, when EDMUND slips away from the Beavers' house to betray his brother and sisters to the WHITE WITCH, LUCY asks, "Can *no one* help us?" and Mr. BEAVER assures her, "Only ASLAN. We must go on and meet him. That's our only chance" (LWW 8). In contrast, as Mrs. BEAVER takes her time even in the face of the likelihood that the witch will get ahead of them, the less-trusting SUSAN asks, "have we no hope?" (LWW 10). In the first instance "chance" could easily be spelled "Hope," whereas Susan is, properly speaking, Hopeless.

In the Narniad Lewis implies uppercase-h hope for the first significant time in PC 14 when CASPIAN's old NURSE says that she has been waiting for Aslan's visit all her life. REEPICHEEP is also a hieroglyph of hope. Caspian says of him, "Reepicheep here has an even higher hope" (VDT 2). Reepicheep makes a PROFESSION OF FAITH as well as hope, "As high as my spirit, though perhaps as small as my stature. . . . I expect to find Aslan's own country. It is always from the east, across the sea that Great Lion comes to us" (VDT 2).

The contrast between hopes and FEARS and real hope (and proper awe before the GREATNESS OF GOD) comes into sharp focus in Prince RILIAN's speech. "[W]hen once a man is launched on such an ADVENTURE as this, he must bid farewell to hopes and fears, otherwise DEATH or deliverance will both come too late to save his HONOR and his reason" (SC 13). Later, when Aslan renews Caspian's life with his own blood, "[a] great hope rose EUSTACE's and JILL's hearts. But Aslan shook his shaggy head. 'No, my dears,' he said. 'When you meet me here again, you will have come to stay' " (16).

LB is the chronicle that most raises and dashes hopes so that real hope, "wild hope" (16), can rise in our hearts. When TIRIAN PRAYS to Aslan, "there began to be a kind of change inside Tirian. Without knowing why, he began to feel a faint hope. And he felt somehow stronger" (4). From this point in the story he and Eustace and Jill hope (lowercase-h) that somehow they can undo the damage that SHIFT and the CALORMENES have done, that they might win the LAST BATTLES, and that Narnia might be saved. When these hopes have been set aside, the wild hope overtakes them: There was a real RAILWAY ACCIDENT, all (save Susan) have died, and they and real Narnia and real ENGLAND have been saved in ASLAN'S COUNTRY—the real STORY has just begun.

[1]FAITH, Hope, and Love are the three theological virtues of both medieval thought and Christian theology. They are virtues (from the Latin word *virtus,* "strength") in that they are powers given to the human mind and the human heart that develop and transfigure the human powers of faith, hope, and love so that human beings can know and trust God (Faith) without seeing God yet, to desire God and the things of God (Hope), and to love God and others self-sacrificially (Love). See Josef Pieper, *Faith Hope Love* (San Francisco: Ignatius Press, 1997). See FRIEND(S), FRIENDSHIP.

[2]Book III, Chapter 10, first ¶.

[3]"Our Lord finds our desires not too strong, but too weak. We are half-hearted creatures, fooling about with drink and sex and ambition when infinite joy is offered us, like an ignorant child who wants to go on making mud pies in a slum because he cannot imagine what is meant by the offer of a holiday at the sea. We are far too easily pleased." "The Weight of Glory" in *The Weight of Glory,* ¶1.

HORN — SUSAN'S gift from FATHER CHRISTMAS, which if blown in great need will always bring help. Its SOUND, richer in tone than a bugle, is first heard when the horn is used to call for help against FENRIS ULF. It is left behind when the CHILDREN returned to ENGLAND at the end of LWW (17). This loss, which seems tragic, is really a PROVIDENCE: By the time of Prince CASPIAN it is the greatest and most sacred treasure of NARNIA, and it is given to the young prince by Doctor CORNELIUS (PC 5). At the end of

PC 15, Caspian offers Susan the horn but she says he may keep it.

HORRORS — EVIL beings, present at the slaying of Aslan (LWW 14).

HORSE(S), HORSEMANSHIP — Horses play a large part in Narnian life and are found throughout the *Chronicles.* The horse, for Lewis, is a symbol of all that is best in having bodies and having natural desires. So good horsemanship, the harmonic interaction of horse and rider, is symbolic of the harmony of our spiritual and physical natures—our bodies give movement to our spirits, which in turn gently guide them.[1] The CENTAUR is a MYTHOLOGICAL picture of this harmony. SHASTA's increasing skills in horsemanship are probably a symbol of the gradual reunion of spirit and nature Lewis speaks of in *Miracles,*[2] whereas ARAVIS's CALORMENE style of riding—rider as total master of the horse—is typical of the Calormene disregard for nature. In BREE and HWIN, the two most fully developed ANIMAL characters in the seven books, the characteristics of horses that every rider knows are brought to their most noble height. That "no one can teach riding so well as a horse" is given its highest meaning when Bree actually tells Shasta how to ride. The acute sense of hearing in horses is acknowledged by Bree's ability to tell by the SOUND of a horse in the distance that it is a thoroughbred mare being ridden by a fine horseman. Lewis also instructs the reader in proper treatment of horses when he notes that galloping day and night is only for STORIES, and that real horses must alternate brisk trots and short walks to stay in the best condition. He also notes the importance of grooming and rubdowns, and the pleasures of water, grass, and hot mash. Of Narnian talking horses in particular we learn that, contrary to Bree's fears, they do love to roll in the grass like other horses, and that although they are not usually mounted they don't mind being ridden on "proper occasions," such as to war.

FRANK, the cabdriver, is always considerate of his horse Strawberry and eventually is crowned the first KING of Narnia. Jadis, the VIOLENT witch who flogs Strawberry unmercifully and maddens him with her whispered words, is eventually killed. The winged FLEDGE carries DIGORY and POLLY to the GARDEN and saves them from the witch. Even the noble JEWEL the UNICORN is in awe of Fledge's beauty.

Other horses mentioned by name in the *Chronicles* include DESTRIER; POMELY; the QUEEN OF UNDERLAND'S horse, Snowflake; and RILIAN'S horse, Coalblack.

[1]C. S. Lewis, *The Great Divorce,* Chapter XI, last 15 ¶¶s.

[2]C. S. Lewis, *Miracles,* Chapter 14, ¶29, and Chapter 16, last 4 ¶¶s.

HOSPITALITY — Freely given hospitality is a fundamental aspect of Narnian life, bespeaking the Narnian delight in DOMESTICITY and attention to COURTESY. Mr. and Mrs. BEAVER open their home to the Pevensie CHILDREN in LWW; TRUMPKIN cares for CASPIAN in his home and differs from NIKABRIK about killing the prince because "it would be like murdering a guest" (PC 5). PATTERTWIG offers nuts, and the CENTAURS share their food. In VDT, CORIAKIN provides a MAGIC meal for LUCY, and the crew and the DUFFERS enjoy a feast. In contrast, CALORMENES demand hospitality. In HHB, ARSHEESH, fearful of reprisal, dares not refuse the TARKAAN. Escaping this Calormene brand of hospitality, SHASTA flees to the north and has his first Narnian meal in the home of the DWARF brothers DUFFLE, Rogin, and Bricklethumb. The GIANTS' hospitality at HARFANG is even worse: They are kind to JILL, EUSTACE, and PUDDLEGLUM only because they intend to eat them at the AUTUMN FEAST.

HOUSE OF PROFESSOR DIGORY KIRKE — A large, complicated country house in the south of ENGLAND located ten miles from the nearest railway station. It is full of all sorts of rooms and places to hide—an armor room, a harp room, a billiard room, rooms and rooms full of books,[1] and one room with only a WARDROBE in it. Old and famous, it is mentioned in guidebooks and histories of the area. In his younger days (MN 15), DIGORY lived in the house with his mother, MABEL, his father, and Uncle ANDREW. How it comes to be destroyed by the end of LB 16 Lewis does not reveal; perhaps it was bombed in World War II.

[1]Lewis's own book-filled boyhood home is lovingly described in his autobiography, *Surprised by Joy*, 10.

HOUSE OF THE MAGICIAN — The home of CORIAKIN is located in the land of the DUFFERS (VDT 13). A long, gray, two-story stone house, grown over with ivy, it sits at the end of a TREE-lined avenue. The house has a long dining hall. The sunlit passage on the second floor that runs the whole length of the house is carpeted, and its walls are carved and paneled. Masks also hang on the wall, and the DOORS are painted with strange undecipherable lettering. A bearded mirror (if you look at it, you can see yourself wearing a beard) hangs just past the sixth door. The last door on the left leads into the room of the MAGICIAN's book. Another room has several mysterious instruments in it—astrolabes, orreries, chronoscopes, poesimeters, choriambuses, and theodolinds.

HUMOR — The *Chronicles of Narnia* are filled with humor from beginning to end. In fact, the CREATION OF NARNIA is no sooner accomplished than the First Joke is made by the JACKDAW, and "all the other ANIMALS began making various queer noises which are their way of laughing and which, of course, no one has ever heard in our world"[1] (MN 10). Then, with ASLAN's permission, they all abandon themselves to merry laughter. The Jackdaw not only *makes* the first joke; he becomes the first joke.[2] Many scenes in the *Chronicles* are reminiscent of music hall routines, the British vaudeville with which Lewis was familiar. Some striking examples include GLIMFEATHER's impossible conversation with the deaf TRUMPKIN; the newly created TALKING BEASTS' discussion of whether Uncle ANDREW is animal, vegetable, or mineral, and their chase of poor Andrew in a reverse foxhunt; the old owl's imitation of Trumpkin at the parliament of owls; and the interplay between the arrogant TISROC, the fawning AHOSHTA* TARKAAN, and the firebrand RABADASH, which moves from burlesque to wry wit.

In the opening epigraph of *The Screwtape Letters,* Lewis quotes Martin Luther as saying, "The best way to drive out the devil, if he will not yield to texts of scripture, is to jeer and flout him, for he cannot bear scorn."[3] In the *Chronicles,*

Lewis employs humor in a similar way. He succeeds in making EVIL figures ridiculous—and thus defuses their power—by poking fun at them, especially Rabadash and Uncle Andrew.

The *Chronicles* have become quite grim by LB, and there is little laughter left in NARNIA. But there is always laughter in ASLAN'S COUNTRY, and at the GREAT REUNION, one of the best features is the retelling of old jokes. Lewis makes his own joke when he adds, "You've no idea how good an old joke sounds when you take it out again after a rest of five or six hundred years" (LB 16).

[1]Lewis says "of course" because animals are grave and solemn in our world and they are thus because they are dumb and witless. Risibility (a term coined by Boethius in the sixth century) is for many philosophers one way of defining the human being: A person is a being with a sense of humor (Aquinas, *Summa Theologica* III, 16, 5). Lewis prized Boethius, calling his *De Consolatione Philosophiae* "one of the most influential books ever written in Latin. . . . To acquire a taste for it is almost to become naturalized in the middle ages." (*The Discarded Image*, 75.)

[2]For a discussion of Lewis's authorial lapse at this point, see JACKDAW.

[3]See also *The Screwtape Letters,* Letter XI.

HUNTING — Hunting ANIMALS for meat is an accepted practice in NARNIA, with one important stipulation: DUMB BEASTS may be killed with as little PAIN as possible; to kill and eat a TALKING BEAST is both murder and cannibalism.

HWIN — The mare ridden by ARAVIS in HHB. Like BREE, she is a Narnian TALKING BEAST who was captured in her early youth and enslaved in CALORMEN. Her name is evocative of the "whinny" sound HORSES make. Although she was necessarily SILENT in Calormen, she is forced to speak in order to stop her mistress's attempted SUICIDE. She will not be silenced by Aravis either: "This is my escape just as much as it is yours," she tells the Tarkheena (2). She is a highly bred mare, "a very nervous and gentle person who was easily put down" (9), but sensible and brave nonetheless. Her stay in the HERMIT's enclosure seems to enhance her COURAGE and enthusiasm, and though she is shaking all over with NUMINOUS* FEAR when she sees ASLAN leap over the hermit's wall, she waits only a short time before she gives "a strange little neigh, and trots across to the Lion." She tells Aslan that he may eat her if he likes: "I'd sooner be eaten by you than fed by anyone else" (14). Aslan, of course, does not eat her, but kisses her and tells her that she shall have JOY. She PRAYS about this experience for two hours. Hwin lives happily ever after in NARNIA, marrying and living to a ripe old age, visiting her friends in ARCHENLAND often.

I

IMAGINATION — Lewis felt very strongly that CHILDREN's imaginations should be nourished and encouraged to grow. Because EDMUND's imagination is stimulated by reading *Robinson Crusoe,* he knows the elements of survival that help keep the children going when they are "shipwrecked" in the ruins of CAIR PARAVEL (PC 1).[1] Of course, imaginations can go bad or interfere with reality. Uncle ANDREW's vain imagination causes him to play down Jadis's fearsomeness, and to exaggerate her beauty and his handsomeness. But when she bursts in on him and LETITIA, her real presence dissipates all of his daydreaming (MN 6). In LB 13, TIRIAN tells DIGGLE that the black hole is all in his imagination, but it becomes a prison real enough to the DWARFS.

In order to encourage children's imaginations, Lewis often turns things topsy-turvy:[2] the books on TUMNUS's bookshelf include *Is Man a Myth?* (LWW 2); CASPIAN suddenly realizes in TRUFFLEHUNTER's cave that Old Narnia really exists (PC 5); Caspian dreams of living on a round world, which he has read about but never believed was real (VDT 15); the porter at the GIANTS' castle at HARFANG speaks to JILL and EUSTACE as if they were beetles and says he supposes they look "quite nice" to one another (SC 7); GOLG is horrified at the thought of living in Overland, with "no roof at all . . . only a horrible great emptiness called sky" (SC 14); SHASTA hears talking HORSES for the first time and BREE refers to Shasta in horse terms, as a "foal" (HHB 1); and TIRIAN becomes aware that he is in a "very queer ADVENTURE" when he notices fruit TREES inside the STABLE (LB).

[1]Lewis no doubt felt that he was helping children with all the PRACTICAL NOTES he included in the *Chronicles.* Lewis almost surely was alluding here to G. K. Chesterton's comment on Robinson Crusoe in *Orthodoxy* (Chapter 4, last 3¶¶s): "Crusoe is a man on a small rock with a few comforts just snatched from the sea: the best thing in the book is simply the list of things saved from the wreck." See LIST(S).

[2]Chesterton might have called this "topsy-turvydom"; see *Orthodoxy,* all of Chapter 4.

Jill and Eustace talk to Puddleglum as the Marsh-wiggle explains—as Marsh-wiggles will—that he is fishing for eels, though he doesn't expect to catch any. (SC 5)

INCUBUSES — Evil spirits of the night, present at the slaying of Aslan (LWW 14). According to medieval thought, an incubus is the personification of the nightmare, a demon who descends on people, especially women, while they sleep.

INSECTS — Narnia is filled with humans, Dwarfs, animals, gods, and trees, but the only insects specifically created by Aslan are butterflies (perhaps for their beauty) and bees (undoubtedly for pollination of the plant life). Eustace and Edmund discuss grasshoppers in the land of the Duffers but decide not to mention them to Lucy, who doesn't like insects—especially large ones.[1] (VDT 10)

[1]Lewis did not like insects either, and that is probably why Narnia is so lacking in them. See *Of Other Worlds*, 30, and Autobiographical allusion(s).

J

JACKDAW — A small bird of the crow family that can be taught to talk. The Jackdaw's gaffe soon after the creation of the Talking Beasts (continuing to talk when the others had fallen silent) is the occasion for the first laughter in Narnia. He truly enjoys the joke at his own expense and asks if everyone will always be told how he made the First Joke. The Jackdaw suggests that Digory and Polly are the Second Joke and labels the bear's pratfall as the Third Joke. (MN 10 and 11)

JADIS — See White Witch.

JEWEL — A creamy-white Unicorn, noble and delicate, with one blue horn in the middle of his forehead and a gold chain around his neck. He is King Tirian's best friend. Soft-spoken and gentle, he is fierce in battle and uses his horn as a weapon. Throughout LB, he steadfastly refuses to believe in the false Aslan, reminding Tirian that Aslan is not a *tame* lion. In Aslan's country at last, he swims up the great waterfall.

JEWELS — See Robe(s), royal.

JILL POLE — She and her schoolmate Eustace Scrubb are the first products of a modern coeducational school system to have Narnian adventures. It is to escape the harassment of the gang at Experiment House that she finds a way into Narnia, and it is partly to be strengthened for returning to this situation that she has been called into Narnia. Her chief failure in Narnia—forgetting the signs Aslan gives her, by which she and Eustace are to find the lost Prince Rilian—is connected to the fact that she is a fearful person and doesn't like physical discomfort. Her time in Narnia in SC (during which her feelings run from delight to depression) prepares her for her role in LB (in which she has matured into a steady, reliable guide).[1]

She and Eustace have a lot to learn about cooperation when they first meet, and they quarrel a good bit in SC. [SC SPOILERS] Not yet aware that one can't force good magic to happen but must ask to enter Narnia, she first proposes to use a darker magic. She and Eustace have scarcely begun their PRAYER to Aslan when they find themselves chased by members of the gang into ASLAN'S COUNTRY. Jill's desire to show off the fact that she isn't afraid of heights leads to Eustace's falling into Narnia.

The most terrible moment Jill has ever faced in life is when she feels all alone at the bottom of the cave they have fallen into. The touch of her companions helps her through the tight spots. Rilian's request of them in the Lion's name stuns her, and she would rather not confront this decision. But when PUDDLEGLUM correctly distinguishes between consequences and commands, she leads the way in releasing the prince. After the witch's death, Jill and Eustace apologize to each other, wish each other well, and use each other's Christian name for the first time.

The reader experiences the pathos of CASPIAN's funeral through Jill. It is she who notices that Eustace is CRYING over the dead king as an ADULT would cry—though it is hard for her to place his age on Aslan's Mountain. She is overwhelmed by tears at Aslan's beauty and by the mournfulness of the funeral music. She and Eustace are sent back to Experiment House as administrators of the Lion's justice upon the gang (she keeps her Narnian finery and wears it to a holiday DANCE). A lifelong FRIENDSHIP with Eustace has begun.

When they return to help TIRIAN in LB [SPOILERS], Jill is sixteen years old and very businesslike in rescuing the king. We see the situation at the STABLE through Jill's eyes. She is moved to tears by the handsome EMETH's youthful zeal for his GOD* TASH. She slays three enemies and wins Tirian's praise; he assigns her to cover for the rest of their company as they make a preemptive strike. And though she feels "terribly alone," she does her duty. She throws herself into the LAST BATTLE and is captured and thrown through the stable DOOR.

When Tirian walks through the same door, he sees a transformed Jill, ageless and beautiful and fresh. She is part of the great procession to meet Aslan for the final scene of the *Chronicles.* From a frighteneded nine-year-old, she has grown to be a loving, fearless queen in the Narnia that lives forever.

[1]She is also the last female character Lewis creates in the *Chronicles,* and her character reveals a Lewis trying very hard to overcome his SEXIST outlook.

JOKE — See HUMOR.

JOY — Lewis uses this word to describe the "wanting" of something or the LONGING for something that is better than the "having" of any particular thing. The GARDEN is "a happy place, but very serious" (MN 13). EMETH describes his joy as a "happiness so great that it even weakens [him] like a wound." The older friends of NARNIA laugh at DIGORY's refrain that "it's all in PLATO," but they become grave again "for, as you know, there is a kind of happiness and wonder that makes you serious" (LB 15).

JUDGMENT — The process in which every living creature must look into ASLAN's eyes. The ones who look upon him with hate or fear pass into his shadow and into oblivion and the ones who love him, even though they are awed, pass through the DOOR into his country (LB 14).

[DEATH; END OF NARNIA; ESCHATOLOGY; HELL.]

K

KEY(S) — Tirian carries a "nice bunch of keys" made of gold, which Lewis carefully describes to remind the reader of the contrast between royal life in untroubled times and the present Narnian crisis (LB 5). At Aslan's command, the High King Peter closes and locks the stable* door with a golden key (LB 14).[1]

[1]This is an allusion both to the popular Christian notion about St. Peter as the keeper of the keys to the gates of heaven (see the Gospel of Matthew 16:19) and to one of Lewis's favorite short stories, "The Golden Key," by George MacDonald.

KIDRASH TARKAAN — The name borne by both the paternal great-grandfather and the father of Aravis. The latter adult is married a second time to a woman who is jealous of Aravis. The stepmother persuades him to give his daughter to Ahoshta* Tarkaan, a wealthy and powerful man whom Aravis despises. When Kidrash realizes that his daughter has fled in order to escape this marriage, he goes to Tashbaan in search of her. (HHB 3)

KING(S), QUEEN(S) — All true kings and queens of Narnia are human beings who rule by the will of Aslan. In MN, Aslan outlines to King Frank I and Queen Helen the five necessary qualities of their VOCATION: (1) to work with their hands to raise their own food; (2) to rule the people[1] as free subjects and fellow human beings; (3) to educate their offspring to rule in the same way; (4) to prefer none among their offspring or their subjects and to tolerate no abuse of one by another; and (5) in the event of war, to be first in battle and last in retreat. Most important, they must rule with humility. In Narnian countries, bad rulers are bad precisely because they lack humility and hold themselves to be above the law. These tyrants and despots include the WHITE WITCH (and her alter ego, Jadis), MIRAZ, the TISROC, and the QUEEN OF UNDERLAND. The named kings and queens of Narnia and ARCHENLAND are Frank I, Helen, Frank V, COL (of Archenland), GALE, OLVIN, PETER, EDMUND, SUSAN, LUCY, LUNE (of Archenland), Cor (of Archenland), ARAVIS (of Archenland), RAM (of Archenland), SWANWHITE, CASPIAN I, CASPIAN VIII, CASPIAN IX, Nain (of Archenland), CASPIAN X, the DAUGHTER OF RAMANDU, RILIAN, ERLIAN, and TIRIAN.

[1]The word "people" as used in this sense includes TALKING BEASTS, DWARFS, and other nonhuman residents of Narnia.

KIRTLE — See QUEEN OF UNDERLAND.

KNIGHT(S), LORD(S), LADY/LADIES, SQUIRE(S) — Mentioned only in passing, they are presumed to be part of the chivalric ORDER and HIERARCHY in NARNIA. Many knights are among the champions who searched for RILIAN and never returned. Rilian himself is said to be a very young knight. Squires and ladies went Maying with the queen and Rilian the day of the queen's death. (SC 4)

KRAKEN — A large, dark, MYTHICAL sea monster, said to dwell in Norwegian waters. Described as one of the most remarkable of all ANIMALS, it belongs, along with the SEA SERPENT and the squid, to the group of three dangerous beasts of the deep feared by the SEA PEOPLE (VDT 15).

L

LADY LILN — See OLVIN.

LADY OF THE GREEN KIRTLE — See QUEEN OF UNDERLAND.

LAMB — In Christian art, a symbol both of innocence and of Christ the Redeemer. Both of these usages appear in the *Chronicles.* At WORLD'S END* EDMUND, EUSTACE, and LUCY encounter a Lamb who is transformed into ASLAN (VDT 16). He reveals that there is a way into his country from earth, that they must enter through this way, that he will always be showing them this way; he does not say how long the way will be, but it always goes across the river of DEATH. He tells them not to fear death because he is the *Pontifex Maximus* (Latin for "greatest bridge builder"), one of the earliest titles for Christ: He has bridged the gap—death—between life and life. In LB 3, the Lamb who observes that NARNIA is Aslan's and CALORMEN is TASH's, and that there should be no FRIENDSHIP between them, is a hieroglyph[1] of innocence.

[Biblical allusion(s).]

[1]See Talking Beasts.

LAMP-POST — Marks the westernmost boundary of Narnia. It grows from a crossbar of the lamp-post outside the Ketterleys' London residence, which is damaged in the crush. Brandished by Jadis as a weapon, it fells two policemen and is transported with her to Narnia. She throws it at Aslan, but it drops to the ground, where it grows into a complete—already lit—lamp-post that shines day and night and gives the name "Lantern Waste" to the region. Encountered first by Lucy, it is rediscovered and marveled at by the royal party while they are on the hunt for the White Stag that brings the Golden Age of Narnia to a close.

[White Witch.]

LANTERN WASTE — The area surrounding the lamp-post, which marks the westernmost boundary of Narnia, takes its name from the eternally lit lamp. It is an uncultivated tract of wilderness, of which Edmund is duke. An ancient forest of talking trees—including the Golden Tree, the silver tree, and the Tree of Protection—has grown up here. Many of the trees are felled in LB. Lantern Waste is mentioned in passing in HHB as within the western limits of the hermit's vision in his pool. Tirian's grandfather had built three protective towers in Lantern Waste[1] to maintain law and order against the robbers who then plagued the area. These towers were simple, functional structures, dark and damp but well stocked with provisions. It is here that Tirian, Jill, and Eustace return after their failed negotiations with the Dwarfs.

[GEOGRAPHY, NARNIAN.]

[1]In 2534 N.Y., according to Lewis's outline of Narnian history (*The Land of Narnia*, 31).

LAPSED BEAR OF STORMNESS — A talking bear who has gone back to the marauding habits of a wild bear, sallying forth from his den on the Narnian side of Stormness Head and attacking travelers. He is reformed by Prince CORIN, who earns the title "Thunder-Fist" when, in the dead of winter, he finds and boxes with the bear for thirty-three rounds (HHB 15).

LASARALEEN TARKHEENA — A childhood friend of ARAVIS. Clothes, parties, and gossip are the top priorities of this model of VANITY. A habitual giggler, she loves to attract attention and travels in a litter "all a-flutter with silken curtains and all a-jingle with silver bells," from which emanate the rich scents of perfumes and flowers. She is married to a great CALORMENE and has a summer home in MEZREEL (VDT 3, 7, 8, and 9). Her name is derived from *lasar*, an obscure Scottish form of *leisure;*[1] the suffix *een* is the Gaelic way to express the feminine.

[1]Martha C. Sammons, *A Guide through Narnia* (Wheaton: Harold Shaw, 1979), 150.

LAST BATTLE(S) — The final struggle of the forces of good (TIRIAN's "army") with the forces of EVIL (RISHDA's CALORMENE cohort) in NARNIA (LB 10). It is in four phases, the final one of which is the "last battle." Tirian has JEWEL, PUZZLE, POGGIN, JILL, EUSTACE, FARSIGHT, fifteen DOGS, MICE, moles, SQUIRRELS, a boar, a bear, and twenty HORSES (in reserve) on his side. Rishda has fifteen Calormene veterans, a bull, a wolf, Slinkey, and WRAGGLE on his side. Aloof from the battle are the renegade DWARFS.

In the first phase (11), the Calormenes attack but are beaten off with the loss of two soldiers, the bull, the fox, and the SATYR. Three dogs are killed and one is wounded. The bear is also killed. The horses, freed by the mice, are killed by the Dwarfs before they can come to Tirian's aid.

In the second phase (11), Tirian's forces sortie but fall back to a white rock when the Calormene troops are reinforced. The Calormene side loses several men and the wolf but Tirian loses Eustace.

In the third phase (12), Rishda divides his force, one group to guard the Narnians at the rock, and the other to attack the Dwarfs who had been shooting at the Calormene soldiers. When those dwarfs who have been taken alive are sacrificed to TASH in the STABLE, Rishda turns his attention to the remaining Narnians.

Thus the final phase (12), "the last battle of the last King of Narnia," begins. The boar is the first to go down. Jill is captured and thrown into the stable DOOR. The last battle ends for Tirian when he pulls Rishda into the door.

The stark COURAGE and the hopeless quality of the Narnian fighters evoke in anyone who has read Norse MYTHOLOGY the clear allusion to *Ragnarok,* translated variously as the last battle of the GODS, the twilight of the gods, or the "destruction of the powers," in which the monsters slay the gods and destroy earth and heaven.[1]

[1]H. R. Ellis Davidson, *Gods and Myths of Northern Europe* (Harmondsworth: Penguin, 1964), 236.

LAST SEA — The body of water between RAMANDU'S ISLAND and WORLD'S END, it lies beyond the known Narnian world (VDT 15). The crew of the *DAWN TREADER* finds, in at least three specific ways, that all is different there: (1) Everyone needs less SLEEP, less food, and less conversation; (2) there is too much light; the SUN at dawning looks two or three times its normal size; and (3) the song of the birds of morning gives a special atmosphere to the place.

The waters of the last sea are SWEET, not salt, a fulfillment of the prophecy of the DRYAD at REEPICHEEP's birth. After a sip of it, CASPIAN drinks deeply and everything about him becomes brighter. He compares the water to light, and Reepicheep calls it "drinkable light." The taste is dazzling, "stronger than WINE and somehow wetter, more liquid, than ordinary water." They can now bear the ever-increasing light and look straight at the growing sun without blinking; everything and everyone is transformed.

LEFAY,[1] MRS. — ANDREW Ketterley's godmother, and according to him "one of the last mortals in this country who had FAIRY blood" (MN 2). Uncle Andrew tells DIGORY that Mrs. Lefay was "queer," and that she did unwise things for which she was put in prison. (To Andrew, "unwise" implies that her deeds were not wrong or unlawful, but that she should have been more careful.) Like Andrew, she didn't associate with "ordinary, ignorant people."[2] Although Uncle Andrew promised her that he would destroy a secret box of MAGIC dust, he does not, turning the dust into RINGS.

[1]Lewis intended to allude to Morgan Le Fay, the enemy of King Arthur; see the "Lefay Fragment" in *Past Watchful Dragons,* 48–67, as mentioned in the *Advice to an Intelligent Reader from an Intelligent Reader,* n. 5.

When Scrubb, Jill, and Puddleglum discover the giants' recipes for cooking Men and Marsh-wiggles, they realize they must flee Harfang before the giant cook wakes up and turns them into dinner. (SC 9)

[2]The adjectives "ordinary" and "ignorant" here are equated in the minds of magicians and technologists who have no sense of RIGHT AND WRONG. See CURIOSITY and TECHNOLOGY.

LEOPARD, THE — A Narnian constellation, seen in the summer sky along with the Ship and the Hammer (LWW 12 and 13; HHB 13; SC 3; MN 9). The Leopard is LUCY's favorite, and it shines on her the night she attempts to awaken the TREES (PC 9). The sight of this trio of "friendly" constellations would have cheered her greatly during her first FEARful night on RAMANDU'S ISLAND (VDT 13).

LEOPARD(S) — See CAT(S).

LETITIA KETTERLEY — DIGORY's Aunt Letty; she is a spinster who lives in London. She has taken in her strange and difficult brother ANDREW and her seriously ill sister MABEL Kirke, Digory's mother.

LILITH — A female demon of both Babylonian and Hebrew MYTHOLOGY, who murders newborn babies, harms women in childbirth, and haunts wildernesses on the lookout for CHILDREN. In LWW 8, she is said to be the wife of a Jinn and mother of the WHITE WITCH (who, true to her heritage, spends most of LWW trying to do away with the Pevensie children).

LILYGLOVES — The funny old chief mole, whom PETER remembers from the old days. Leaning on his spade, he says that someday King Peter will be glad of the newly planted apple orchard (PC 2).

LIST(S) — Lewis's love of the ordinary[1] shows up in many ways in the *Chronicles:* his attention to the details of SMELLS, TACTILE IMAGES, SOUNDS, DOMESTICITY, and especially lists. In LWW 7, he lists the items to be found in the BEAVERS' house and the menu for the Beavers' meal, as well as the happy ANIMAL-person sounds as the statues in the courtyard of the castle of the White Witch come alive (16). In PC 2, he lists the treasures of CAIR PARAVEL and the foods eaten at the feast after the war with MIRAZ (15). In VDT 4, Lewis outlines the provisions in the *DAWN TREADER;* the things on GUMPAS's table that come down in a cascade; the things worth having (including beer, WINE, timber, cabbages, and books); the menu for the DUFFERS' hearty supper (10); the menu of Lucy's MAGIC luncheon with CORIAKIN and the real and imaginary instruments in the HOUSE OF THE MAGICIAN (11); and the menu of the continuous banquet on RAMANDU'S ISLAND (13). In HHB 4, the description of TASHBAAN from a distance and its crowded streets is accomplished by lists, as is the menu for SHASTA's CALORMENE (5) and Narnian (12) meals. In SC 3, the menu for supper at Cair Paravel is given.

[1]See IMAGINATION, n. 1.

LITERARY ALLUSION(S)[1] — Lewis was known as one of the best-read persons of his time, and he shares his great fund of knowledge with the readers of NARNIA.

In PC 15, Lewis calls REEPICHEEP "the Master Mouse," a term suitable enough to Reepicheep's station in life but actually drawn from Henryson's version of Aesop's lion-and-mouse story in the *Morall Fabillis.* No doubt Lewis was tickled by the expression and amused by the new currency he was giving it. He may even have looked ahead with pleasure to the possibility that a few of his child readers, going on to study literature, would meet Henryson's original Master Mouse. Children might sooner recognize PUDDLEGLUM's testing of the bottle by degrees as an allusion to Pooh's testing of the honey-jar in the Heffalump Trap (SC 7). Lewis may also be trying to educate modern children, as Professor Kirke often echoes Lewis's own feelings: "What do they teach them in school nowadays?" Thus in VDT 16, EDMUND infuriates CASPIAN by comparing Reepicheep's plan to tie him up with the crew's binding of Ulysses to the mast so that he could do nothing beyond hearing the sirens' song. Lewis perhaps hoped that children might be curious enough to find out for themselves what this Ulysses story was all about. Other literary allusions are noted in the entries to which they apply.

[All allusions are spelled out in the text or first note of the following entries, except where noted: ALBATROSS; ANDREW KETTERLY; ASLAN'S VOICE; BACCHUS; BASTABLES; BIBLICAL ALLUSION(S); CORIAKIN; DEPLORABLE WORD, THE; DESTRIER; DUFFER(S); EARTHMEN; FATHER TIME; FRANK I, n. 2; GARDEN; GEOGRAPHY, NARNIAN; GIANT(S); GREED; IMAGINATION; JEWEL; KEY(S); LEFAY, MRS.; NARNIA; OGRE(S); ORKNIES; PATTERTWIG; PAVENDERS; PETER PEVENSIE, n. 3; PITTENCREAM; POMELY; POMONA; PRIZZLE, MISS; PRUNAPRISMIA; PUZZLE; REEPICHEEP; SALAMANDERS OF BISM; SEA SERPENT; SHERLOCK HOLMES; SHIFT; SOUND(S); *SPLENDOUR HYALINE;* STABLE; STARS; STONE KNIFE; TISROC; TRUFFLEHUNTER; WARDROBE; WOOD BETWEEN THE WORLDS; WOOSES.]

[1]The source of many of those allusions is Professor T. W. Craik, University of Durham, England.

LONGING — The term—along with JOY—Lewis uses to express the sort of experience within life that opens us up beyond what we see to the GREATNESS OF GOD and the goodness of everything else, the craving suggested but not fulfilled by pleasure. Longing is often manifested as HOPE. In LWW 14, the girls have longed ever since they first saw him to bury their hands in ASLAN's mane. In PC, the young Prince CASPIAN longs for Old Narnia (4), and Doctor CORNELIUS has been looking for traces of it all his life (5). At the sight of the Narnian sunset, JILL is filled with a longing for ADVENTURE (SC 3). In HHB 1, both BREE and SHASTA long to go to NARNIA, Bree because he remembers how free he was as a colt, and Shasta because he has had a somewhat mysterious desire to go there all his life.

The pleasant MEMORY just out of reach is also an expression of longing. Thus the TELMARINE soldier in PC 15 has a startled, happy look, "as if he were trying to remember something"; and FRANK, POLLY, and DIGORY, entranced by the joy of Aslan's song of creation, look "like they are being reminded of something." In LB 12 and 15, Aslan is referred to as TIRIAN's "heart's desire"; EMETH's desire since boyhood has been to serve and know TASH and to look upon his face; and JEWEL says of ASLAN'S COUNTRY, "This is the land I have been looking for all my life, though I never knew it till now"

[RIGHT AND WRONG.]

LUCK — See PROVIDENCE.

LUCY[1] PEVENSIE — The youngest CHILD and second daughter of Mr. and Mrs. PEVENSIE; she is known as Queen

Lucy the Valiant in the GOLDEN AGE OF NARNIA, is one of ASLAN'S closest friends, and is perhaps the best-developed character in the *Chronicles.* She is the person through whom the reader sees and experiences most of NARNIA in LWW, PC, and VDT; she is absent from only SC and MN, is in the background of HHB, and figures in the last quarter of LB. She is a fair-haired, happy, and compassionate person, deeply sensitive and intuitive, but somewhat fearful and vain. Her STORY is one of growth from FEAR to COURAGE so that she becomes known as Lucy the Valiant (LWW 17). It is through her character that Lewis expresses his own religious sensibilities. Through her, as well, the reader sees the connection between the first hearing of ASLAN'S NAME (at the SOUND of which she has a beginning-of-vacation, waking-up-in-the-morning FEELING) and the last words of Aslan ("The term is over: the holidays have begun. The DREAM is ended: this is the morning"). From beginning to end, she is concerned about other people: Her first question of Aslan is to ask him to do something about her brother EDMUND; she pleads for mercy for RABADASH; almost her last question is to ask the Lion to try to help the renegade DWARFS. She is the one who notices how terrible Aslan's paws are and the one to experience their playfulness and caresses. He shares with her his laughter and happiness as well as his sadness. It is part of Edmund's PROFESSION OF FAITH that Lucy sees Aslan most often. To her is given the gift of healing in the CORDIAL made from the juice of the SUN'S FIRE-BERRIES. Her capacity for instant FRIENDSHIP endears her to TUMNUS, ARAVIS, REEPICHEEP, her brothers, and even the mysterious sea-maiden.

In LWW, she is between eight and nine years old. All

the charms of the new world of Narnia come to the reader through her lively senses. She is swept up in Aslan's explanation of the meaning of his resurrection, so much so that she does not know why she is laughing as she scrambles over the top of the broken STONE TABLE to reach him. The one fault she commits in LWW is her desire to see the results of her cordial upon the wounded Edmund. She is cross with Aslan but responds immediately to his growl both with words of contrition and with actions on behalf of the other wounded. And it is she who notices the depth of Edmund's transformation.

Lucy is the first in PC to feel the tug from Narnia. Chapter 9, "What Lucy Saw," emphasizes her gifts of insight. Her special kind of wakefulness is an ineffable signal to go out to meet the TREES and attempt to awaken them. She has a momentary vision of Aslan that her companions ought to trust but don't; though she is angry at their contrariness, she very generously does not criticize them. Aslan favors her with an ecstatic, nighttime rendezvous, but he does not let her complain about her companions—he means only for her to strike out on her own if they do not listen to her. This is a hard lesson for her to learn and she would rather escape it in the intimacy of their relationship, but this intimacy itself strengthens her. It is very difficult for her to hear the grumbling of her companions, but she forgets her harsh responses by keeping her attention focused on the Lion who walks ahead of her. She is compensated for this hardship by the many DANCES that take place in Narnia as Aslan liberates the country and its people. Her sorrow that PETER and SUSAN will not be able to return to Narnia is genuine; she is a very selfless person.

[VDT SPOILERS] Lucy cares for EUSTACE even before he has come to the point of being able to think of anyone besides himself: She uses her cordial to heal his seasickness and shares her water with him during the rationing. Even before she knows he is the DRAGON, she eases its pain with her cordial and kisses it. She has only to hear of the need of the invisible people before she volunteers to seek out the MAGIC spell for them. In the scene at the magician's book, Lucy's faults become most evident. She is insecure: She thinks herself inferior to her sister, especially in physical beauty; and she worries about how devoted her friends are—she wants very much to be liked. So strong is her desire to be accepted that when she is prevented from reciting the spell that would make her irresistibly beautiful, she rushes to recite the very next spell that would have a bearing on this desire—the spell to know what her friends think about her. But hearing her own voice raised in anger at her friend MARJORIE Preston's betrayal reminds her that she is talking only to a picture and not really to her friend. She then recites the spell for the refreshment of spirit, which is such a lovely story that she wants to read it over again but finds that she cannot have an encore: The pages only turn forward. She can remember only a few main images from the story. Lucy finally finds and reads the spell to make hidden things visible. Aware that she has made *all* things visible, she turns to face the open DOORS of the library (their being opened had made her somewhat uncomfortable when she first came into the room). She is transfigured by what she sees—"Her face lit up till, for a moment (but of course she did not know it), she looked almost as beautiful as that other Lucy." What she sees is Aslan.

Though she feels otherwise, she tells CASPIAN she is willing to sail into the dark. The reader senses the complete horror of the DARK ISLAND from her vantage point in the fighting top. She hangs on tight and whispers a PRAYER (the first explicit prayer in the *Chronicles*) for Aslan's help.

The wonders of the LAST SEA are, for the reader, largely Lucy's experience. She becomes absorbed in trying to understand the life and culture of the SEA PEOPLE, whose country she sees beneath the clear waves, and she experiences instant FRIENDSHIP with the Sea Girl. When Lucy tastes the SWEET waters, she gasps that this is the loveliest thing she has ever tasted, but so very strong and nourishing that she will need to eat no further. The waters so strengthen her that (though she feels some pangs at leaving the *DAWN TREADER* behind at WORLD'S END) she is too excited by the light, the SILENCE, the SMELL, and even the special kind of loneliness of the place to mourn too long over the parting.

Lucy asks the LAMB they see if this is the way to ASLAN'S COUNTRY. As he tells her no, he is transfigured into the Lion. As he moves to show them the door to their own country, Lucy begs him to let them return soon. Hearing his "never again," she cries out his name in despair. She sobs her most beautiful lines in all of the *Chronicles:* "It isn't Narnia, you know. It's you. We shan't meet you there. And how can we live, never meeting you?"

[LB SPOILERS] In LB, Lucy accompanies POLLY, DIGORY, JILL, and Eustace on the train that will meet Peter and Edmund. After the RAILWAY ACCIDENT, she finds herself dressed as a queen and standing with her friends in the middle of a discussion about the meaning of a STABLE whose door alone she can see. She does not speak until the end of this conversation; when she does, she speaks with a thrill of JOY in her voice that communicates itself immediately to TIRIAN ("She was drinking everything in more deeply than the others. She had been too happy to speak"). Now the reader not only sees with her eyes but understands the meaning of everything through her understanding. She tries to befriend the renegade DWARFS, but to no avail. After flying up the falls and across the west to the GARDEN, the SEVEN FRIENDS OF NARNIA all meet Reepicheep, then Tumnus, and all their friends. Lucy sits with Tumnus (her first Narnian friend) on the wall of the garden overlooking Narnia and again tries to comprehend what has happened to them. Through Lucy's eyes, the reader sees Aslan bounding toward them. Hers are the final words addressed to him in the *Chronicles;* she speaks their fear that they will be sent back into their world. He removes this fear forever and announces the real, everlasting ADVENTURES further up and in his country.

[1]When Lewis began writing LWW in 1939, he wrote, "This book is about four children whose names were Ann, Martin, Rose, and Peter. But it is mostly about Peter who was the youngest." In honor of his goddaughter, Lucy Barfield, he dedicated the book to her and changed its main character to a youngest child named Lucy (see DEDICATION[S]).

LUNE — KING of ARCHENLAND, father of twins Cor (Shasta) and CORIN. That he is a good and fair ruler is shown over and over: He prevents EDMUND from killing RABADASH by declaring the CALORMENE unfit for a gentleman's death because he has breached the code of HONOR and war-COURTESY; he reproves his son Corin for joining the battle but only thinly disguises his pleasure over his son's COURAGE; he was merciful to Lord BAR, his chancellor, when he discovered that Bar was an embezzler (however, when he found out that Bar was also a Calormene spy and had kidnapped the king's son, he pursued and killed him); he explains to Cor that he must be king, for this is his inescapable duty and VOCATION. An unpretentious man, he greets ARAVIS not in his royal ROBES but in his oldest clothes. (HHB 5, 11, 13, 14, and 15; LB 16)

M

MABEL KIRKE — Mother of DIGORY Kirke, younger sister of ANDREW Ketterley and LETITIA Ketterley. For most of MN (1, 2, 6, and 9–15) she is seriously ill—Uncle Andrew uses her illness as a means of coercion against Digory, and its cure by the Apple of Youth is seen as a miracle. She seems to be a kind ADULT who sings and plays the piano and enjoys playing games with Digory and POLLY. She is very strict about keeping promises, and has taught Digory not to steal—a virtue that serves him well when he is tempted in the GARDEN.

MACREADY, MRS. — Housekeeper to Professor DIGORY Kirke (LWW 1, 5, 6, and 17). She is not fond of CHILDREN and dislikes being interrupted when she is telling visitors all she knows about the house. A stereotype of the British housekeeper, she is bossy and full of instructions.

When the children return from NARNIA, she is still talking with the visitors outside the spare room—an indication of the wide discrepancy between English TIME and Narnian time. Mrs. Macready's name is probably based on that of Mrs. McCreedy, a woman who kept house for Lewis's parents when he was two years old,[1] although it may also be a play on the words "make ready."

[1]Kathryn Lindskoog, *Voyage to Narnia: Response Book* (Elgin, IL: David C. Cook, 1978), 13. See AUTOBIOGRAPHICAL ALLUSION(S).

MAENADS — In Greek and Roman MYTHOLOGY, the female members of BACCHUS's company. They leap, rush, turn somersaults, and dance the DANCE of plenty (PC 14 and 15).[1]

[1]For the meaning of their cry, *Euan, euan, euoi-oi-oi-oi,* see BACCHUS, n. 1.

MAGIC, MAGICIAN — Even without the popularity of J. K. Rowling's wonderful Harry Potter books, it would be important for the *Pocket Companion to Narnia* to devote space to a fairly complete discussion of C. S. Lewis's thoughts about magic (black magic and white magic, EVIL magic and good magic). In his autobiography, Lewis says that at the time of his mother's illness and death he had been told "that PRAYERS offered in FAITH would be granted."

> I accordingly set myself to produce by will power a firm belief that my prayers for her recovery would be successful; and, as I thought, I achieved it. When nevertheless she died I shifted my ground and worked myself into a belief that there would be a miracle. . . . I had approached God, or my idea of God, without love,

> without awe, even without fear. He was, in my mental picture of this miracle, to appear neither as Savior nor as Judge, but merely as a magician; and when He had done what was required of Him I supposed He would simply—well, go away.[1]

The nine-year-old CHILD Lewis confused faith with FEELINGS of faith. He was operating with a definition of *magic* as "power to do whatever one asked" and of *magician* as "one who has such power." But to do all possible things belongs to God and not to human beings. Lewis knew that ADULTS, YOUTH, and children want to have power over their lives and over nature—these desires are not RIGHT OR WRONG in themselves. But evil happens when humans seek this power "at times, or in ways, or in degrees, which God has forbidden."[2]

In the last year of his life a sixty-four-year-old Lewis wrote about good magic in the book that was in the press when he died, *Letters to Malcolm, Chiefly on Prayer.* In the following sentences taken from Letter XIX, he discusses the "magic" of what some Christians call the Eucharist and others Holy Communion or the Lord's Supper:

> Here [in Holy Communion] a hand from the hidden country touches not only my soul but my body. Here the prig, the don, the modern in me have no privilege over the savage or the child. Here is big medicine and strong magic. . . . (103) ¶When I say "magic" I am not thinking of the paltry and pathetic techniques by which fools attempt and quacks pretend to control nature. I mean rather what is suggested by fairy-tale sentences like "This is a magic flower, and if you carry it

> the seven gates will open to you of their own accord," or "This is a magic cave and those who enter it will renew their youth." I should define magic in this sense as "objective efficacy which cannot be further analysed." (103) ¶Now the value, for me, of the magical element in Christianity is this. It is a permanent witness that the heavenly realm, certainly no less than the natural universe and perhaps very much more, is a realm of objective facts. (104)

In the *Chronicles,* "magic" at first calls to mind various evil enchanters, enchantments, and instruments we encounter in NARNIA: the petrifying WAND of the WHITE WITCH in LWW; TURKISH DELIGHT; the spell of the Hundred Years of Winter that lies over Narnia; NIKABRIK's willingness in PC to tap dark magical forces, if necessary, to overthrow MIRAZ; the spell of forgetfulness cast over Prince RILIAN by the QUEEN OF UNDERLAND in SC; the DEPLORABLE WORD of Empress Jadis of CHARN; and Uncle ANDREW's magic RINGS.

But the WARDROBE that lets the children into Narnia is also magic, good magic. Narnia knows good magicians such as CORIAKIN and Doctor CORNELIUS. The Deep Magic, which the witch counts on to ensure her triumph over ASLAN, is not quite what she takes it to be—its real dimensions include a Deeper Magic that undoes all her evil. In short, the *Chronicles* do not present a battle between the ordinary, natural world and dark, unnatural magical powers. The ordinary, natural world is rather the battlefield on which the good magic confronts the bad, just as ultimately the human heart is the stage on which the mystery of grace wrestles with the mystery of evil.

The *Chronicles of Narnia* are themselves a kind of magic, a seven-volume magician's book devised by Lewis to break bad enchantments of TECHNOLOGY and VANITY and CURIOSITY and bring about re-enchantments by means of HONOR, COURAGE, and OBEDIENCE by reawakening a LONGING for Aslan and ASLAN'S COUNTRY. Lewis spoke in "The Weight of Glory": "Do you think I am trying to weave a spell? Perhaps I am; but remember your fairy tales. Spells are used for breaking enchantments as well as for inducing them. And you and I have need of the strongest spell that can be found to wake us from the evil enchantment of worldliness which has been laid upon us for nearly a hundred years."[3] Lewis was well aware of Tolkien's insight: "Small wonder that *spell* means both a story told, and a formula of power over living men."[4]

Deep Magic — A complex term in LWW used by the White Witch and also well known to Aslan, expressing the demands of justice (LWW 13). Since it has existed "from the dawn of time," it is not eternal; it will last only as long as worlds outside of Aslan's country last. The words of the Deep Magic are inscribed in the three sacred places: on the STONE TABLE (a place of ritual sacrifice); on the firestones of the SECRET HILL (another sacrificial context), or (only in the American editions) on the trunk of the WORLD ASH TREE (with its Odin imagery of inescapable destiny); and on the SCEPTRE of the EMPEROR-BEYOND-THE-SEA, the strongest support of its claims. The witch understands it only to the extent that she knows that a traitor's blood (i.e., life) belongs to her. Aslan's knowledge of it is far deeper and he will not work against it. Rather, in obedience to it and to the Deeper Magic, he offers his own life in EDMUND's place.

Deeper Magic — A complex term in LWW, the full meaning of which Aslan alone knows (LWW 15); it is something like self-sacrificing love. The Deeper Magic is eternal (the meaning of "from before the dawn of time"). That it is not simply "mercy" in contrast to the "justice" of the Deep Magic is clear from Aslan's definition of a willing and innocent victim substituting himself or herself for a traitor, whereupon DEATH is undone and the need for sacrifice in order to fulfill the Deep Magic is abolished. This process finds echoes in the way a volunteer must sail to Aslan's country at the WORLD'S END in order for the enchantment of the THREE SLEEPERS to be broken (VDT 14), and in the spirit of TIRIAN'S offer to let himself be killed so that Narnia might be saved (LB 4).

[1]*Surprised by Joy*, 20–21.

[2]At the end of paragraph 2 of Letter IX of *The Screwtape Letters*, the tempter laments: "All we can do is to encourage the humans to take the pleasures which [God] has produced, at times, or in ways, or in degrees, which He has forbidden."

[3]"The Weight of Glory," *The Weight of Glory*, ¶5.

[4]"On Fairy-Stories," ¶39 (this essay first appeared in *Essays Presented to Charles Williams*).

MAN-HEADED BULL — A MYTHOLOGICAL creature, enthusiastically on ASLAN's side in the war against the WHITE WITCH (LWW 12 and 13). He is not to be confused with any of the MINOTAURS, bull-headed men who are invariably on the witch's side.

MARCH(ES) — The border districts of NARNIA. King EDMUND is count of the Western March, west of LANTERN WASTE. The unnamed HERMIT lives in the Southern March, which borders on ARCHENLAND. RILIAN rides in Narnia's Northern Marches in search of the serpent that killed his mother. JILL, EUSTACE, and PUDDLEGLUM are taken captive by the WARDEN OF THE MARCHES OF UNDERLAND; he is a servant of the QUEEN OF UNDERLAND.

MARJORIE PRESTON — An intimate, older school chum of LUCY Pevensie (VDT 10). Lucy violates Marjorie's PRIVACY when she "overhears" Marjorie betray her during the eavesdropping enchantment. In order to gain ANNE FEATHERSTONE's respect, Marjorie asserts that she doesn't care for Lucy as much as she appears to. Lucy, surprised and hurt, asserts that Marjorie, who is apparently the kind of girl who has few FRIENDS, should be grateful for all that Lucy has done for her.

MARSH-WIGGLE(S) — In SC, the humanlike beings that live in marshes and subsist on a diet of eels and a strong, alcoholic beverage. The derivation of the name is apparent: "marsh" from their habitat and "wiggle" from their rather rubbery physiques and affinity for eels. They refer to themselves as "wiggles." Their feet are webbed and cold-blooded, "like a duck's."[1] They value their PRIVACY, and live in wigwams[2] spaced far apart. They SMOKE a heavy tobacco, perhaps mixed with mud. Internally, they are quite different from human beings—although in what way is not specified. However, PUDDLEGLUM's hand, chewed by GOLG the EARTHMAN, would have been a bloody pulp were he not a Marsh-wiggle. Marsh-wiggles do the Narnian work that is concerned with water and fish, and JILL and EUSTACE are ferried across the Fords of Beruna by a ferry-wiggle. Marsh-wiggles are on duty at CAIR PARAVEL and fasten the hawsers when ships dock there. The general disposition of Marsh-wiggles is dour and pragmatic, and they generally expect the worst of any situation. GLIMFEATHER suggests them as the best guides to take Jill and Eustace across ETTINSMOOR.

[1]A mistake: Ducks are warm-blooded.

[2]Perhaps an abbreviation, in this case, for "wiggle-wams."

MAUGRIM — See FENRIS ULF.

MAVRAMORN — One of the SEVEN NOBLE LORDS and one of the four TELMARINE visitors to the Island of the Voices in the years 2299–2300 N.Y. He is one of the THREE SLEEPERS at ASLAN'S TABLE (VDT 2, 11, and 13).

MAZERS — Hardwood drinking bowls used at the feast celebrating the victory of Old NARNIA over MIRAZ and the TELMARINES (PC 15). Mazers are cups or goblets without feet, which were originally made out of mazer wood (wood derived from gnarled tree burls) and often richly carved or

The Queen of Underland transforms herself into a hideous, seething serpent and coils herself around Rilian. Scrubb and Puddleglum come to his rescue with swords at the ready while Jill looks on in horror and tries not to faint. (SC 12)

ornamented with silver and gold or other metals. The "great wooden bowl, curiously carved" mentioned in LB 2 could be a mazer.

MEMORY — Lewis felt that one's memory of an experience actually improves that experience and might, indeed, be *better* than the original. In PC 4, LUCY has a thrill of memory as she recognizes three of NARNIA's summer constellations. When Lucy tries to remember the STORY that is the spell for refreshment of the spirit in VDT 10, it fades on her: She cannot have it again because the pages turn only forward, not back. However, she experiences *reminders* of it for the rest of her life in other stories that she reads, and she judges a story as good to the extent that it "reminds her of the forgotten story in the MAGICIAN'S book." EUSTACE is concerned that he will remember to his dishonor how he passed up the "marvelous ADVENTURE of exploring Bism" (SC 14). Perhaps the most vivid memory in the *Chronicles* is TIRIAN'S memory of his childhood, which comes flooding back with his father, ERLIAN'S, embrace (LB 16).

Lewis also stresses the importance of trying to remember. JILL thinks she should say something courteous to ASLAN when he finishes giving her the four SIGNS, so she says, "Thank you very much. I see" (SC 2). Aslan, who really *does* see, gently says that she may not really see as well as she thinks she does and tells her that the first step toward seeing is remembering—and the first step toward remembering is repeating. He commands her four times to remember the signs, adding the fourth time that she should also believe them. He adds that, in the thicker air of Narnia, it may be hard to remember her QUEST, and that unless she learns the signs by heart it will be easy to forget them.

MERMEN, MERMAID(S) — Inhabitants of the Eastern Sea on its border with NARNIA. They sing at the coronation of PETER, EDMUND, SUSAN, and LUCY (LWW 17 and PC 2). Their unforgettable music differs from ordinary music in its strangeness, SWEETNESS, and profoundly piercing character. Lucy confuses them with the SEA PEOPLE of the LAST SEA (as do many readers), but Edmund reminds her that the mer-people are able to live in air and in water, whereas the Sea People live only in water (VDT 15).[1]

[1]When writing LWW, Lewis probably did not know he would create in VDT a separate species, called by the proper name "Sea People." Hence he uses the phrase generally in LWW.

MEZREEL — A summer resort of the wealthy in CALORMEN (HHB 3). Its attractions include a lake, famous gardens, and the Valley of the Thousand Perfumes—an area of especially fragrant flowers and blossoming TREES and shrubs. To LASARALEEN Tarkheena, whose summer home is here, Mezreel is the epitome of sensual delights.

MICE — Mice figure prominently in the *Chronicles*. They were not created as TALKING BEASTS, but because they were kind enough to gnaw through the ropes to free the dead ASLAN (LWW 15), they were given the GIFT OF SPEECH (PC 15). Ever helpful, mice rally to TIRIAN's side and are sent to free the HORSES in LB 4 and 11. They are over two feet tall and stand on their hind legs. The most famous mouse is the courageous REEPICHEEP.

MINOTAURS — Enormous, earthshaking men with bulls' heads, totally devoted to the cause of the WHITE WITCH (LWW 13, 14, and 15). They are present at the slaying of ASLAN and take an active role in the battle with Aslan's armies. They are not to be confused with the MAN-HEADED BULL who fights on Aslan's side. In Greek MYTHOLOGY the Minotaur was the Cretan monster who, housed in the labyrinth, devoured his annual tribute of seven Athenian youths and seven Athenian maidens until he was slain by Theseus.

MIRAZ — The TELMARINE* KING of NARNIA, Prince CASPIAN's uncle, and the husband of PRUNAPRISMIA (PC 4, 5, 7, and 13). Affectionate to Caspian only out of necessity, he is hoping for an heir of his own. He is almost a stereotypical ADULT who does not believe in fairy tales, the possibility of talking ANIMALS, or anything that he sees as impractical. Miraz himself is a usurper. Calling himself the lord protector, he weeded out his opponents, including the SEVEN NOBLE LORDS, through HUNTING "accidents," dangerous assignments, and false charges of treason and madness. [PC SPOILERS] When his own son is born he decides to murder Caspian; he spends the rest of PC tracking him down. Lords GLOZELLE and SOPESPIAN flatter Miraz into accepting a MONOMACHY by King PETER. He fights well but trips over a tussock and is stabbed to death by Glozelle.

MONOMACHY — PETER uses this term in his challenge to MIRAZ (PC 13). It means single combat (one fighter from each side representing all the forces of his or her side), an ancient form of battle often used in order to decide the outcome of a war without resorting to general bloodshed. REEPICHEEP takes an interest in the conversation of the predragoned, preconverted EUSTACE only because the boy demands of CASPIAN that he be allowed to "lodge a disposition" with the nearest British consul; the mouse thinks lodging a disposition might be a new form of single combat (VDT 2). Later, Reepicheep desires to challenge the DRAGON Eustace to a monomachy (VDT 6).

MOONWOOD — A Narnian talking hare with hearing so acute that he is able to hear what is whispered in CAIR PARAVEL from his home below the GREAT WATERFALL. His is one of the UNFINISHED TALES (LB 8). According to Lewis's outline of Narnian history,[1] Moonwood lived around 570 N.Y.

[1]*The Land of Narnia*, 31.

MULLUGUTHERUM — The only named EARTHMAN-in-waiting to the QUEEN OF UNDERLAND. It is probable that he is the "prying chamberlain" who helps to strap RILIAN into the SILVER CHAIR. He may also be one of the two gnomes who bow to the queen as she enters the prince's apartments (SC 10 and 11). The SOUND of his name is suggestive of the fawning nature of his personality.

MYTHOLOGY, MYTHOLOGICAL — A myth is a story about what might have actually taken place in history or one explaining why things are the way they are. C. S. Lewis wrote that Christianity is founded on the great myth of the dying and rising god who has somehow communicated a new life to humanity, with the distinction that the

Christian myth had become fact in the historical Jesus of Nazareth. Lewis believed that FAITH is based both on the facts and also on the seeing, hearing, touching, tasting, and smelling elements that give meaning to those facts. This is the point Lewis is trying to make in the dialogue between RAMANDU and EUSTACE about what a STAR is (VDT 14):

> "In our world," said Eustace, "a star is a huge ball of flaming gas."
>
> "Even in your world, my son, that is not what a star is but only what it is made of."

You can reason about or *look at* someone else's experience all day and be able only to get the facts about it; it is only when you *look along* that person's experience (if not actually, at least in a story) that you can see, hear, touch, taste, smell the meaning of what that person is experiencing.

Reading the *Chronicles*, you will encounter mythological beings from many of the world's ancient cultures, especially Greek and Roman. In NARNIA you find BACCHUS, CENTAURS, DRAGONS, DRYADS, DWARFS, FAUNS, FENRIS ULF, GIANTS, HAMADRYADS, the KRAKEN, MAENADS, the MAN-HEADED BULL, MINOTAURS, NAIADS, NYMPHS, the PHOENIX, POMONA, the RIVER-GOD, SATYRS, SILENUS, SILVANS, and UNICORNS. These creatures spring forth naturally as a response to ASLAN's command to Narnia to awake and become divine. In the foreground of the Narnian tapestry are the grave and playful characters from classical Greek and Roman mythology, the Fauns and satyrs and Dryads who bring a spirit of REVELRY to Narnia. Lewis has woven the background in the darker colors and less distinguishable patterns of the stark and terrifying characters from the north, the witches and giants and Black Dwarfs of Celtic, Norse, and old German mythology.

In the *Chronicles* Lewis emphasizes that STORIES often contain facts and truths. Prince CASPIAN consoles himself in MIRAZ's castle by DREAMING of the tales his NURSE has told him of Old Narnia, which the TELMARINES consider to be fairy tales. It is his belief in these stories, and the wisdom they impart, that saves him from the same fate that befell his ancestors and allows him to become a true KING of Narnia in the tradition of the High King PETER. His realization in TRUFFLEHUNTER's house that the stories are true is a NUMINOUS experience that Lewis wants his readers to share.

NAIAD(S) — In Greek and Roman MYTHOLOGY, water NYMPHS. In the *Chronicles,* Naiads are one of the NINE CLASSES OF NARNIAN CREATURES, and one of NARNIA's names is the country of "Visible Naiads" (PC 4). The four daughters of the RIVER-GOD emerge with him from the GREAT RIVER at ASLAN's command to the waters of Narnia to be divine (MN 10); they are Queen HELEN's trainbearers at her coronation (MN 14). The young Prince CASPIAN calls them "nice people" who used to live in the streams of Narnia (PC 4); since the coming of the TELMARINES and their pollution of the rivers, however, the Naiads have fallen into a deep sleep (PC 6). In LWW 12, Lewis mentions Well-Women who play stringed instruments and are frightened by the wolf attack, and it is probable that they are Naiads.

NARNIA — The enchanted world of Lewis's *Chronicles;* also the name of the country that is the northern neighbor of ARCHENLAND. The CREATION OF NARNIA is accomplished through ASLAN's song, and the END OF NARNIA is also at his bidding. The HISTORY of Narnia lasts 2555 Narnian years (N.Y.), which corresponds to forty-nine earthly years (A.D. 1900–1949).[1] Its GEOGRAPHY is as varied as our own; the world of Narnia has deserts and moors, oceans and rivers, mountains and valleys. Politically, it is composed of three kingdoms: Narnia and Archenland to the north of the great desert, and CALORMEN to the south. The history of Narnia is largely uneventful, consisting of long stretches of peace punctuated by wars and invasions. ASLAN'S COUNTRY exists outside of Narnia; indeed, it is not physically connected to any country.

Old Narnia — The physical world created by Aslan in MN; characterized by the presence of all manner of fauna and flora that, at Aslan's command, become divine and alive. Old Narnia is meant to be ruled by humans, who do so with humility. It is a country where COURTESY, HONOR, and the spirit of ADVENTURE are the respected virtues, and in which HOSPITALITY is graciously offered.[2] Old Narnia exists before, during, and after the rule of the New Narnians. This world is destroyed by the DRAGONS and dinosaurs at Aslan's command.

Old Narnian(s) — Specifically, the nonhuman residents of Narnia: DUMB BEASTS and TALKING BEASTS, DWARFS and GIANTS, GODS and goddesses, creatures of MYTHOLOGY, and wild people of the wood. During the rule of the New Narnians they are SILENCED and forced into hiding, and many of the wood and water spirits go to sleep to await the WAR OF DELIVERANCE.

New Narnia, Real Narnia — The heart of Narnia, which has always existed in Aslan's country, on the other side of the DOOR. From the GARDEN, the CHILDREN behold Narnia, which they have just seen destroyed. But what was destroyed was the Old Narnia, the Shadow Lands;[3] the renewed Narnia—the real Narnia—exists externally in Aslan's country.

New Narnian(s) — The TELMARINES; specifically, the dynasties of CASPIAN I through MIRAZ. "New Narnians" is a term distinct from "New Narnia." The rule of CASPIAN X restores Old Narnian rule.

[ASTRONOMY, NARNIAN.]

[1]See *The Land of Narnia*, 31

[2]For a comparison of Narnian and Calormene ways, see CALORMEN.

[3]See PLATO.

NIKABRIK — A Black DWARF in PC 5 and 12 [SPOILERS], the second voice CASPIAN hears upon regaining consciousness. From his first appearance he shows the nature "soured by hate" that Caspian later attributes to him. Naturally suspicious, he is for killing Caspian immediately but is prevented by TRUMPKIN and TRUFFLEHUNTER. He would also do away with the half-dwarf Doctor CORNELIUS, whom he sees as a renegade. But his greatest aggravation is Trufflehunter, the faithful badger. Nikabrik is not only negative; he is a downright disbeliever. When asked if he believes in ASLAN, he answers that he believes in *anyone,* be it Aslan or the WHITE WITCH, who will liberate NARNIA. This equating of Aslan's power with that of the witch intensifies his confrontation with Trufflehunter, who sees the witch as Narnia's worst enemy. Nikabrik protests that she was good to the Dwarfs, and for him this is all that matters: Dwarf interests come first. He has no use for the simple pleasures of pipe SMOKING, and he does not join in the DANCE before the great council. He is the very antithesis of COURTESY.

NINE CLASSES OF NARNIAN CREATURES — According to Doctor CORNELIUS, they are Waking TREES, Visible NAIADS, FAUNS, SATYRS, DWARFS, GIANTS, GODS, CENTAURS, and TALKING BEASTS. He adds that Old Narnia is not the land of men (PC 4).[1]

[1]Lewis's handling of humans is inconsistent. Later in the *Chronicles* he indicates that humans have been present in Narnia since the beginning. One indication of this is the place Lewis assigns to King FRANK and Queen HELEN (MN 11).

NUMINOUS — In the *Chronicles,* most characters experience the numinous[1]—awe and delight, fear and attraction, felt at the same time in the presence of something holy—in ASLAN's presence. Their first sight of Aslan in LWW 7 cures the CHILDREN of the mistaken notion (held by all who have not been to NARNIA, says Lewis) that a thing cannot be good and terrible at the same time. CASPIAN has a numinous experience when he realizes that all the STORIES about Old Narnia are true; he feels terror at the thought that the DWARF is "not a man at all" and delight that "there are real Dwarfs still, and I've seen one at last" (PC 4). When Aslan turns to face the children and TRUMPKIN, he looks so majestic that "they feel as glad as anyone can who feels afraid, and as afraid as anyone can who feels glad" (PC 11). At LUCY's first sight of the new constellations in VDT 13, she is filled "with a mixture of JOY and FEAR." At the SOUND of ASLAN'S VOICE, JILL's fear turns to awe and she is "frightened in rather a different way," that is, with a fear of the numinous (SC 2). SHASTA's encounter with the Lion in HHB 11 begins with the fear of GIANTS and GHOSTS and ends in the "revelation" of the TRINITY and Shasta's "new and different sort of trembling," which is also a feeling of gladness. In MN 11, DIGORY finds Aslan to be "bigger and more beautiful and more brightly golden and more terrible" than he had thought. At the END OF NARNIA the creatures who pass through the STABLE* DOOR and look into Aslan's face with both love and fear move to the right into ASLAN'S COUNTRY (LB 14). EMETH experiences the numinous in the extravagant CALORMENE fashion, comparing the terror and beauty of Aslan to the FLAMING MOUNTAIN OF LAGOUR and to a rose in bloom (LB 15).

[HWIN; LONGING.]

[1]For Lewis's views on the numinous, see *The Problem of Pain,* Chapter 1, ¶5 ff. See also Rudolf Otto, *The Idea of the Holy* (London: Oxford University Press, 1968), 31 ff.

NURSE — The beloved woman who takes care of Prince CASPIAN as a CHILD, to whom she tells tales of Old Narnia (PC 4, 5, and 14). MIRAZ, displeased with her STORIES, banishes her from the castle without even a chance to say good-bye, and Caspian misses her greatly. (She also incurs the wrath of NIKABRIK for having told Caspian about Old Narnia.) Later in Caspian's ADVENTURES, near Beaversdam, ASLAN finds a little old woman near DEATH who looks as if she might be part DWARF. She makes her final peace, professing her lifelong FAITH and trust. Then, calling Aslan by name, she asks if he has come to take her away. Instead, he makes her well. She appears to be modeled on Lewis's beloved nurse, Lizzie Endicott, who introduced him to the stories of Beatrix Potter and to Irish folktales.

[AUTOBIOGRAPHICAL ALLUSION(S); BIBLICAL ALLUSION(S).]

NYMPHS — In Greek and Roman MYTHOLOGY, beautiful semidivine maidens who live in a variety of natural habitats—TREES, rivers, mountains, and so on. In LWW 2, they are mentioned as DANCING with FAUNS at midnight festivals. In PC 11, they rise out of the river at ASLAN's roar. In MN 14, four river-nymphs—the NAIAD daughters of the RIVER-GOD—bear Queen HELEN's train. The sons of King FRANK and Queen Helen marry nymphs (MN 15).

O

OATH(S) — Narnian oaths and promises are to be taken seriously, and several oaths are specifically mentioned in the *Chronicles.* All KINGS and queens of NARNIA must swear a coronation oath (REEPICHEEP reminds CASPIAN of his solemn oath to serve Narnia in order to dissuade him from his planned abdication in VDT 16). In SC 6, Lewis reinforces a cardinal moral tenet for CHILDREN when he says that the children couldn't tell without PUDDLEGLUM's consent, because they had promised. Breaking promises can lead to great EVIL.

OBEDIENCE, OBEDIENT — The word *obedience* is from the Latin *oboedire,* which means "to listen," and it is in this very specific sense that Lewis uses the word in the *Chronicles.* The first response of the newly created TALKING BEASTS is to pledge their obedience to ASLAN with the words "We hear and obey" (MN 10). Phony obedience is found in the CALORMENE class system, where everyone answers those of a higher status with "To hear is to obey." Obedience to Aslan is not something that must be enforced; the Narnian knows the STOCK RESPONSE instinctively. For example, PETER doesn't feel brave when he anticipates fighting FENRIS ULF, "but that made no difference to what he had to do" (LWW 12).

A part of obedience is humility: accepting and living in the truth of obedience, as when BREE has to learn to be "nobody very special" (HHB 10). When Aslan asks CASPIAN if he feels sufficient to rule NARNIA, Caspian demurs, citing his YOUTH and inexperience. Aslan approves of his humility, saying, "If you had felt yourself sufficient, it would have been proof that you were not" (PC 15). RABADASH, on the other hand, is infuriated to the point of tears by King LUNE's treatment of him as a traitor. Lewis comments, "he couldn't bear being made ridiculous. In TASHBAAN every one had always taken him seriously" (HHB 15). In Aslan's presence, the good Narnian must have humility; DIGORY forgets his fame among the Narnians because he is looking at Aslan (MN 14).

OCTESIAN — One of the SEVEN NOBLE LORDS (VDT 2). His device, a hammer and a star,[1] marks the arm-ring found on the DRAGON* EUSTACE. It is not certain whether he became a dragon or was devoured by one, but he did meet his end on DRAGON ISLAND (VDT 7).

[1]Lewis may have patterned Octesian's device on Durin's Emblem, the hammer and anvil crowned with seven stars, from J. R. R. Tolkien's *The Lord of the Rings* (Boston: Houghton Mifflin Co., 1965), Second Edition, vol. I (*The Fellowship of the Ring*), 318.

OGRE(S) — Man-eating monsters, the size of GIANTS, summoned by the WHITE WITCH to the slaying of ASLAN (LWW 13, 14, and 17). They are eagerly present for the murder, and one of them shears off Aslan's mane with its "monstrous teeth." At least three are present at the following day's battle, for EDMUND kills that many. Ogres seem to remain in NARNIA, for the FIVE BLACK DWARFS are in favor of enlisting an ogre or two to overthrow MIRAZ (PC 6). *Ogre* is a French word related to *Orcus,* the Latin name for the god of the underworld; readers of Tolkien will note the significance of this.

OLVIN — The fair-haired KING of ARCHENLAND who in 407 N.Y.[1] defeats Pire, the two-headed southern GIANT, and turns him into stone (thus *Mount* Pire, the two-forked mountain). As the result of this exploit the Lady Liln becomes his wife and queen. Their STORY becomes the subject of many a minstrel's song and an UNFINISHED TALE (HHB 13).

[1]*The Land of Narnia,* 31.

ONE'S OWN STORY — See Privacy.

ORDERS, CHIVALRIC — The code of chivalry is especially important to Narnian knights, ladies, and squires, and the lack of chivalry on the part of the Calormenes and Telmarines is one of their great flaws. Two orders are mentioned in the *Chronicles* (PC 13): the Most Noble Order of the Lion, named, of course, after Aslan; and the Noble Order of the Table, after the Stone Table of Aslan's sacrifice. Peter confers the Most Noble Order of the Lion on Caspian, and Caspian confers the order on Trufflehunter, Trumpkin, and Reepicheep (PC 15).

ORKNIES — Monsters, present at the slaying of Aslan on the Stone Table (LWW 14). *Beowulf* (line 112) mentions "orcheas" among other monsters, and Lewis is content to make a literary allusion to them in order to add to an atmosphere of horror surrounding Aslan's death. The allusion to Tolkien's *orcs* is inescapable.

ORRERIES — See Coriakin.

P

PAIN, HUMAN AND ANIMAL — C. S. Lewis took human and animal pain extremely seriously.

Pain was part of Lewis's autobiography (as it is of every life): He suffered from physical and emotional pain.

He hated going to the dentist but knew that the dentist needed to inflict some pain in order to heal his frequent toothaches. He was wounded in World War I and suffered from AGING AND DISABILITY. Lewis also experienced emotional pain, perhaps the greatest of which was at the DEATH of his mother from cancer when he was nine. He agonized over his brother Warren's absence during World War II and struggle with alcoholism. He went through all the pains of his childhood again, as well as the pain of a husband and stepfather, when he married a woman dying of cancer who would soon leave two sons bereft of a mother.

In spite of these experiences, Lewis believed that the possibility of pain has to exist in a world like ours of humans and ANIMALS; otherwise, one human or animal would not know where it/she/he ended and another human or animal began. When we bump into one another in one way or another, sometimes pain is the result. For example, when a person steps on a cat's tail, or a cat scratches a dog that enters its territory, or a person falls off his or her bike, the cat and the person experience pain. But when a human being inflicts pain on another human or animal or allows such a being to experience any unnecessary pain, this is a moral EVIL.

The problem of animal pain was one that Lewis pondered for at least a quarter century.[1] Cruelty to animals—even vivisection, the practice of experimenting on living animals in order to gain scientific, especially medical, knowledge—was for him a dangerous moral area that led necessarily to the cheapening of human suffering. Lewis had no use for CURIOSITY, the immoderate striving for KNOWLEDGE that sometime corrupts the world of TECHNOLOGY.

Thus EUSTACE would have become a vivisector (he liked to pin dead insects on a card [VDT 1]) had he not been changed in NARNIA. Evidence of this change is seen in his confronting CARTER, one of the GANG of bullies at EXPERIMENT HOUSE, for having tortured a rabbit (SC 1). What is potential in Eustace becomes actual in ANDREW Ketterley, whose cruel experiments with animals had caused him to fear and hate them. Thus he casually excuses his experiments with the MAGICAL dust on guinea pigs with the statement that they are his because he bought them; that's what they're for. He mocks DIGORY's implicit suggestion that he ought to have asked the guinea pigs' permission before he used them and concludes, "No great wisdom can be reached without sacrifice" (MN 2). The only sacrifice he is prepared to make, however, is of POLLY. For this disregard of the sacredness of life, he is punished by being the subject of the "experiments" of the newly created TALKING BEASTS (MN 11). The worst of cruelty to animals is seen in the WHITE WITCH's treatment of Strawberry (MN 7), her reindeer, the DWARF, and Narnian Talking Beasts (LWW 11), and in the CALORMENES' flogging of talking HORSES in LB 2. Vivisection is the very antithesis to proper Narnian relations with animals, talking and dumb.

[1]An animal lover himself, Lewis insisted on getting beneath the emotional arguments on either side of the question. His basic position was that vivisection was a procedure to be used only in extreme circumstances. He summarizes it well: "If on grounds of our real, divinely ordained superiority [human life over animal life] a Christian pathologist thinks it right to vivisect, and does so with scrupulous care to avoid the least dram or scruple of unnecessary pain, in a trembling awe at the responsibility he assumes, and with a vivid sense of the high mode in which

As the four Overlanders—Puddleglum, Jill, Eustace, and Rilian—look on, the gnomes rejoice at the news of the wicked Queen of Underland's death. (SC 14)

human life must be lived if it is to justify the sacrifices made for it, then (whether we agree with him or not) we can respect his point of view." From "Vivisection," 226, in *God in the Dock*. See also "The Pains of Animals" in that volume. A Christian cookbook would need to "exclude dishes whose preparation involves . . . unnecessary animal suffering" (*Christian Reflections*, 1).

PASSARIDS — A family of great TELMARINE lords in the reign of CASPIAN IX (PC 5). They are sent by MIRAZ to fight GIANTS on the northern frontier, where they fall one by one.

PATTERTWIG — A magnificent red squirrel,[1] he is a chatterer (he gives a new meaning to "TALKING BEAST"). Though NIKABRIK thinks squirrels to be generally flighty, he considers Pattertwig trustworthy enough to go to LANTERN WASTE to await the help summoned by SUSAN'S HORN (PC 6 and 7).

[1]In the "LEFAY Fragment," an early draft of MN (see *Advice to an Intelligent Reader from an Intelligent Reader*, n. 5), Pattertwig is met by DIGORY. The red squirrel offers Digory a nut and some advice, especially about Grey Squirrels, who are almost always EVIL (*Past Watchful Dragons*, 50–52).

PAVENDERS — The delectable and beautiful rainbow-colored saltwater fish served at the royal banquets at CAIR PARAVEL (PC 3 and SC 4).

PEEPICEEK — One of the MICE in REEPICHEEP'S company. He is devoted to his leader and vouches that all the mice are prepared to cut off their tails if Reepicheep must go without his (PC 15). Reepicheep designates Peepiceek as his successor (VDT 14).

PEOPLE OF THE TOADSTOOLS — Evil spirits, probably responsible for the poisonousness of toadstools. The White Witch summons them to the Stone Table (LWW 13).

PETER PEVENSIE — The oldest child and son of Mr. and Mrs. Pevensie; Sir Peter Fenris-Bane[1] and High King Peter the Magnificent; victor over the Northern giants in the Golden Age of Narnia; Aslan's helper in renewing Narnia in PC; a student preparing for university entrance examinations with his tutor, Professor Kirke, in VDT; and the one who consigns Tash to his "own place" in the name of Aslan and his Father in LB. He is the figure of the fine older brother to his brother and sisters, and the model king in Narnia.[2] With very few setbacks, he grows from a thirteen-year-old boy in LWW to a splendid, twenty-seven-year-old king in HHB and a twenty-two-year-old university student with his heart still in Narnia when he meets his death in the railway accident that sends the seven friends of Narnia and the Pevensie parents into Aslan's country forever in LB.

From the fact that he is interested in eagles, stags, and hawks in the first pages of LWW, the reader can discern that Peter is predisposed to love and rule Narnia. After Lucy claims to have visited Narnia, he wonders if his sister is well; but the professor challenges him to find her a liar, or crazy, or telling the truth—the only alternatives.[3] Peter provisionally decides on the latter but reserves judgment until he has more data. Once in Narnia himself, he thinks that Lucy ought to be the leader. He feels obligated to try to rescue Tumnus and defends their guide, the robin, by citing the fact that in the STORIES he has read they are always good birds. By this enumeration Lewis means to suggest that Peter is a natural leader, discerning and well read and imbued with the right STOCK RESPONSES.

When Peter first hears Aslan's name, his response is to feel brave and adventurous, two qualities he will need in the years ahead. Mr. BEAVER praises Peter's reaction to the thought of meeting Aslan: "I long to see him, though I am thoroughly frightened"—a classic NUMINOUS experience. From FATHER CHRISTMAS he receives Aslan's gifts of a shield, a sword—RHINDON—and sheath and sword belt with the SILENCE and solemnity called for at such an occasion. After some prodding from SUSAN, he leads his group forward to meet the Lion. A proper sense of responsibility makes Peter acknowledge his fault in EDMUND's treachery—he had stayed angry with his brother instead of seeking to be reconciled with him immediately.[4] Peter is also properly silent after Aslan shows him the country he is to rule, but he goes into immediate action at the SOUND of Susan's HORN. Ignoring his feelings of FEAR, which make him nauseated, he assails the wolf and kills him. Afterward, he feels more tired than brave; he kisses and cries with Susan, two emotional displays that are noble in Narnia. That he should always clean his sword is one of many PRACTICAL NOTES Lewis makes throughout the *Chronicles*. Peter welcomes the traitor Edmund back. When it appears that he might have to lead the fight against the witch's forces without Aslan's help, he worries through the night. After the battle, Lucy sees that his face has become pale and stern and that he seems much more mature. Peter shows his selflessness by giving Edmund credit for the success of the first attack. LWW ends with the tall, deep-chested king urging his royal companions on in their HUNT for the WHITE STAG.

[1]In the British editions, "Wolf's-Bane." Lewis made this change at the same time he changed the name "Maugrim" to "Fenris Ulf" in the first American edition.

[2]Doctor CORNELIUS tells CASPIAN that if he can be a king like Peter, he will have lived up to the VOCATION of being king.

[3]Lewis uses the very same argument numerous times in discussions about the claims of Jesus Christ to be the Son of God. See, for example, *Mere Christianity*, Book II, Chapter 3, "The Shocking Alternative," ¶¶8–11. It is an argument that might have been suggested to him by G. K. Chesterton's *Orthodoxy*, Chapter II, "The Maniac," ¶¶14–15.

[4]Recall the Gospel of Matthew 5:21–24. And notice how in other scenes of JUDGMENT before Aslan, other children have to be led to confess their misdeeds (e.g., JILL and DIGORY); Peter is exceptional in his immediate candor.

PEVENSIE, MR. AND MRS. — The parents of PETER, SUSAN, EDMUND, and LUCY Pevensie. Although they are hardly mentioned at all, they seem to be good and thoughtful ADULTS, because in LWW 1 they have sent the CHILDREN from their London home to the country HOUSE OF PROFESSOR KIRKE to escape the air raids of World War II. Mr. Pevensie is also a professor, and in the summer of 1942 (VDT 1) he and his wife take a sixteen-week trip to America, where he has been invited to lecture. The journey is Mrs. Pevensie's first real vacation in ten years, and she takes Susan along. In 1949 (LB 13), the Pevensies are on a train to Bristol—providentially, it is the same train that Lucy, Edmund, Peter, POLLY, DIGORY, EUSTACE, and JILL are on—when the train is wrecked. From the GARDEN in ASLAN'S

COUNTRY Peter, Edmund, and Lucy see their mother and father waving to them from what seems to be ENGLAND. Aslan explains to them that they have all been killed in the RAILWAY ACCIDENT, and that since the real England is connected to Aslan's country, a short walk will take them to their parents (LB 16).

PHOENIX — A symbol of resurrection, the Phoenix is a MYTHOLOGICAL bird that periodically burns itself to DEATH on a funeral pyre and is reborn from the ashes. It watches DIGORY in the GARDEN in MN 13, and it sits in the TREE above the thrones of King FRANK and Queen HELEN, surveying the GREAT REUNION (LB 16).

PIRE, MOUNT PIRE — See OLVIN.

PITTENCREAM — A cowardly milquetoast (his name suggests someone who is pitiful or a pittance) and the only sailor on the *DAWN TREADER* whom CASPIAN does not allow to go to WORLD'S END (VDT 14). FEARful of such a journey, Pittencream is even more afraid of being left behind on his own. He stays on RAMANDU'S ISLAND, remorseful at not having gone on with the others, and finds his situation intolerable. He boards the *Dawn Treader* on its return voyage but feels so left out, not having shared any of the crew's recent ADVENTURES, that he deserts ship at the Lone Islands and goes to live in CALORMEN. There he tells tall tales of his voyage to the end of the world until he comes to believe them himself. Lewis's final comment, "So you may say, in a sense, that he lived happily ever after," suggests Pittencream's mediocrity and inability to deal with reality.

And the final irony that he could never bear MICE confirms his COWARDICE, since the chief mouse REEPICHEEP is a hieroglyph of COURAGE.

PLATO — A Greek philosopher (427?–347 B.C.), student of the philosopher Socrates and teacher of the philosopher Aristotle. Plato believed that earthly things are material copies of spiritual ideas. He is famous for using (1) the Socratic method of questioning and for his (2) Allegory of the Cave.

(1) In LWW, the Pevensie CHILDREN, troubled over LUCY'S STORY of visiting another world, come to the professor for advice. As Socrates used to do, the professor replies not with answers but with questions that invite the children to step beyond their opinions about what is and what is not possible. He steers the children into a searching and thoughtful openness. When SUSAN protests that "all this about the wood and the FAUN" couldn't be true, the professor replies—as Socrates often does—"That is more than I know." When PETER says that the others looked into the WARDROBE, but *they* did not find NARNIA, the professor challenges their assumption ("[I]f things are real, they're there all the time"—"Are they?"). The contrast between appearance and reality is a related and central Platonic theme.

The QUEST for reality and the clash of appearance and reality are also themes in PC. As in LWW, Lucy has an experience that the others will share only later. The others would rather trust the appearances than Lucy's testimony.[1] Lucy, the apparent DREAMER among her more sober-minded brothers and sister, is the one who is really wide-awake, just as Plato portrays Socrates as the one wide-awake man in ancient Athens.

(2) Chapter 12 of SC is a variation on Plato's famous Allegory of the Cave. What appears to the cave-dwellers on the wall in front of them seems to be real, but it is only the flickering shadows cast by more original objects held up against a fire behind the backs of the cave-dwellers. Freedom comes when one of the cave-dwellers is forced to turn around and see what is behind his back and behind the wall-appearances. He then ascends out of the cave into the open world above and sees that the artificial objects and the fire within the cave were themselves only inferior copies of yet more original realities—real living beings and the SUN. In SC, the witch seeks to halt the ascent of RILIAN, PUDDLEGLUM, JILL, and EUSTACE out into the Overworld. Her tactic is to argue that they can't return to Narnia because there is no Narnia to return to. The queen is the archetypical REDUCTIONIST, trying to convince them that the shadows of their immediate experience are all there is. In trying to defend their belief in the Overworld, the children liken the sun of that realm to the lamp in the witch's apartment, and ASLAN to a huge CAT with a mane. The witch counters that the lamp and the cat are the "realities" and the sun and Aslan are glorified, make-believe "copies." The effect of her discourse, her music, and her MAGIC powder is to induce forgetfulness and weakness of thought. The witch almost succeeds because "the more enchanted you get, the more certain you feel that you are not enchanted at all."

DIGORY's remark "It's all in Plato" (LB 15) refers to Plato's original-copy idea. For Plato, everything that we see in the world around us is a copy of something in a higher, more perfect, spiritual reality. If you were able to go "further up and further in," you would find not something utterly new but something strangely familiar, something "like" what you had always known before, only supremely better. Digory explains to Peter that the Narnia that has passed away was but the shadow of the real Narnia, "which has always been here and always will be here: just as our own world, ENGLAND and all, is only a shadow or copy of something in Aslan's real world." And Lucy need not mourn over the Narnia that was but is no more, because everything that mattered in old Narnia has come home to the new.

[1]The famous Aristotelian saying that "all men want to know" (in the opening line of his *Metaphysics*) must always be balanced by Plato's dialogues. All people, says Plato, also have a desire *not* to know, a tendency to resist the truth, a tendency toward self-deception. For a discussion of the self-imposed blindness of the renegade Dwarfs, see IMAGINATION and DWARF(S).

POESIMETERS — See CORIAKIN.

POGGIN — A helpful DWARF in LB 7, 8, 11, 12, 14, and 16 who leaves the renegade Dwarfs and joins TIRIAN's forces at the LAST BATTLE, putting himself in the KING's and ASLAN's service. Ultimately, he comes into ASLAN'S COUNTRY and is greeted by the SEVEN FRIENDS OF NARNIA.

POLLY PLUMMER — The first human CHILD[1] to enter the WOOD BETWEEN THE WORLDS by means of the MAGIC* RINGS. She is a very balanced person, a good judge of character, perceptive, imaginative, and filled with the spirit of ADVENTURE and creativity. When she appears for the last time in LB—forty-nine years after her first appearance—she is unmarried and is addressed as Lady Polly.[2]

From the contents of her smugglers' cave (a refuge she has constructed in the attic of her house), by her cashbox full of personal treasures, and by the copybook that contains a story she is writing, you can tell that she is an independent Edwardian girl. She is also brave, prudent, rational, and a real leader. She is not tempted by the enchanted GOLDEN BELL and HAMMER's invitation to "know what would have happened."[3] At the first sound of the Lion's creation song she is transported with JOY, another indication of how good she is. And when she sees the Lion, she falls instantly in loving awe of him, the proper STOCK RESPONSE.

[LB SPOILERS] Polly and DIGORY become very good friends, and she hosts a banquet for the SEVEN FRIENDS OF NARNIA. Lewis emphasizes her "wise, merry, twinkling eyes." She dies in the RAILWAY ACCIDENT, on her way with her friends to obtain the magic rings from PETER and EDMUND for JILL and EUSTACE to use to get back into NARNIA. She finds herself restored to youthfulness on the other side of a STABLE* DOOR. Her last words in LB are "Do you remember?" as the sights and SMELLS awaken in her the MEMORY and the HOPE that she has arrived at the goal of her LONGING. This hope is confirmed.

[1]And one of the first humans to enter Narnia. If Lewis intended that the TELMARINES were earthly pirates from the nineteenth century or before, then they would have arrived in Narnia before Polly, but only by means of a wrinkle in TIME. It is clear that Lewis did not harmonize his account of human presence in Narnia in the first three chronicles (by means of the Telmarine incursion through a MAGIC cave on an island in our earth's South

Seas) with his account in the last four books (by means of FRANK and HELEN, as well as Digory and Polly). It is possible that Lewis liked his later explanation better than his earlier and, had he lived longer, would have explained the Telmarines differently.

[2]A sign of her status in Narnia rather than a title.

[3]For the emptiness in knowing what would have happened, see POSITIVITY.

POMELY — Lord GLOZELLE'S HORSE (PC 13). "Pomely" is an adaptation of the old French word for "dappled" and is a LITERARY ALLUSION to the Reeve's horse in Chaucer's "Prologue" to *The Canterbury Tales.*

POMONA — In Roman MYTHOLOGY, the GODdess of GARDENS and fruit TREES. PETER says she put good spells on the apple orchard (PC 2).

POSITIVITY — A term that denotes Lewis's feeling that what has actually happened is most important; and that it is futile to speculate about what might have happened had another choice been made. In PC 10, ASLAN tells LUCY that nobody is ever told what would have happened if . . . In VDT 10, the pages of the MAGICIAN'S book cannot be turned back for a rereading of the spell for the refreshment of the spirit. When he has become visible, Aslan reminds Lucy that no one is ever told what would have happened. In MN 4, the verse on the pillar taunts DIGORY with the idea that not knowing what would have happened will drive him mad, but POLLY, as the voice of reason, states that it doesn't matter what would have happened. However, in MN 14, Aslan tells Digory what would have happened if someone had stolen a silver apple for NARNIA, and what would have happened if he had taken the apple directly to his mother. But because he did the RIGHT thing, these circumstances did not occur.[1]

[1]See RIGHT AND WRONG and PROVIDENCE.

POWER — See MAGIC; TECHNOLOGY.

PRACTICAL NOTES — Lewis's use of LISTS, his love of DOMESTICITY, and his inclusion of practical notes (beginning with not shutting the WARDROBE door in LWW 1) help to make the fantasy world of NARNIA quite real. A great lover of walks and the outdoors, Lewis was not fond of camping. He managed instead to walk from inn to inn, where he could sit down to a hot meal and a fresh bed. Thus in PC 3 he notes the discomfort of sleeping outside and the difficulty of eating roasted apples and hot fish without forks. From his great experience as a walker, he can comment about how easy it is to find imaginary paths in the woods (9). Helpful to the modern reader is the fact that swords aren't very useful for cutting ropes, because they can't be held anywhere lower than the hilt (9). TRUMPKIN'S problem in the rowboat (where his feet don't reach the floor) is a problem common to all CHILDREN in an ADULT world (9). LUCY knows that the best way of getting to sleep is to stop trying (9). In VDT 1, Lewis's great love of boats comes into play, and he notifies readers that in order to read the book intelligently, they must be able to distinguish *port* from *starboard*. When thrown into deep water, Lucy kicks off her shoes—as anyone should to keep from drowning (VDT 1). In SC 3, Lewis comments that, like other deaf

Queen Jadis hangs on in rage as the magic rings transport Digory and Polly out of Charn and into the timeless Wood between the Worlds. (MN 5)

people, Trumpkin is not a "good judge of his own voice." And when EUSTACE loses his temper at PUDDLEGLUM, Lewis remarks parenthetically that people who have been frightened often do (SC 5). Just as it is difficult to see after gazing into a bright light, the travelers can't see after gazing into the bright depths of Bism (SC 14). In HHB, the reader learns all sorts of practical information about HORSES, including the fact that they shouldn't sleep in their saddles (9). And at the enclosure of the HERMIT OF THE SOUTHERN MARCH, Lewis comments that goat's milk has a rather shocking taste to those not used to it (HHB 10).

PRAYER(S) — As discussed in AUTOBIOGRAPHICAL ALLUSION(S), the failure of GOD to answer Lewis's prayers at the time of his mother's illness and DEATH marked the beginning of Lewis's atheism, which lasted until his early thirties. It is remarkable, then, that he is known among Christians as one of the greatest teachers of the power and practice of prayer. His basic teaching:

> In prayer we allow ourselves to be known as persons before God. In prayer we unveil before God. We learn, first, to tolerate and, then, to welcome God's loving gaze and touch. Prayer is, then, personal contact between incomplete persons and God; in this contact he shows himself to us and, un-self-aware, we become persons. This process begins when we show God who we are now and what we honestly want. Whatever desires we have must be the subject of our prayers. Petition, penitence, thanksgiving, and adoration are the traditional four forms in which this growth as persons

> through personal contact with God takes place. Prayer is the beginning of heaven.[1]

If these are the definitions/descriptions we bring to our reading of the *Chronicles,* then nearly every interaction between any human CHILD or TALKING BEAST and ASLAN can be seen as prayer in some fashion. The most illustrative are LUCY's prayer to Aslan for help and his answer in VDT 12; Aslan's words to JILL in SC 2, "You would not have called to me unless I had been calling you"; DIGORY's tears and Aslan's tears (MN 12); and TIRIAN's cry to Aslan for help and the consequences of that cry in LB 4.

[1]From Paul F. Ford, "Soul-Befriending, The Legacy of C. S. Lewis," Spirituality (Dublin), vol. VI, no. 33 (November–December 2000): 357–361.

PRIVACY — Respect for the privacy of "one's own STORY" is important in NARNIA. Lewis himself had an intense sense of privacy, and this theme is apparent in the *Chronicles*. In LWW 13, after EDMUND is rescued from the WHITE WITCH, ASLAN tells him things that no one ever hears and then returns him to the others, saying, "Here is your brother, and—there is no need to talk to him about what is past." In VDT 7, the reformed EUSTACE does not pry when Edmund tells him that he himself was a traitor on *his* first time in Narnia. In the land of the DUFFERS, LUCY comes across a spell in the MAGICIAN's book that enables her to know what her friends think of her. As a result of this invasion of their privacy, she will never be able to forget what they said (VDT 10). RAMANDU quells CASPIAN's CURIOSITY by telling him "it is not for . . . a SON OF ADAM to know what faults a STAR can commit" (VDT 14). In HHB, when SHASTA asks the Lion why he wounded ARAVIS, Aslan replies, "Child, I am telling you your story, not hers. I tell no one any story but his own" (HHB 11). When Aravis asks what harm might come to the slave she drugged, Aslan again replies, "I am telling you your story, not hers" (HHB 14). This is Lewis's doctrine of privacy, and it is not to be broken.[1] The GARDEN in the west is obviously private: "Only a fool would dream of going in unless he had been sent there on a very special business" (MN 13). Finally, arousing the curiosity of many readers, Aslan's words to PUZZLE in LB 16 are for the DONKEY's ears alone.

[1]Cor wants his story told to LUNE, but when Aravis does so he finds that it's not as enjoyable as he thought it would be—in fact, he tires of Lune's constant retellings (HHB 15). This may be one of the punishments for violating the rule prohibiting personal storytelling.

PRIZZLE, MISS — The schoolmistress at the Beruna girls' SCHOOL (PC 14). She reprimands GWENDOLEN for looking out the window at ASLAN's holiday party and accuses her of talking nonsense when she claims to see a lion. When Miss Prizzle herself sees Aslan, she flees, taking her "dumpy, prim little girls" with her. The name "Prizzle" may be a LITERARY ALLUSION to Miss Prism, the governess of Cecily Cardew in *The Importance of Being Earnest* by Oscar Wilde.

PROFESSION(S) OF FAITH — To speak of professions of FAITH in the *Chronicles* is to try to discern whether there are any LISTS or formulas that contain the basic convictions that Narnians believe about ASLAN and his ways of working with them. Mr. and Mrs. Beaver (their prophecies throughout LWW 8 and their sense of Aslan's comings and goings in LWW 17), Trufflehunter (PC 5), and Puddleglum (SC 12) speak snatches of such formulas. The clearest example of a Narnian profession of faith is in Edmund's answer to Eustace's questions in VDT 7: Who is Aslan? and Do you know him? Edmund gives a sixfold reply:

> *"Well, he knows me."* This indicates that the important thing is to be known by Aslan and also that a comprehensive knowledge of him and his ways is impossible to achieve.
>
> *"He is the great Lion, the son of the Emperor-over-the-Sea."* This indicates his basic attributes: his greatness and his relationship to his mysterious Father.
>
> *". . .who SAVED me and saved NARNIA."* This is the first use of the term "saved" in the *Chronicles.* Aslan fulfilled the demands of the Deep Magic with respect to Edmund's treachery, and the Lion also delivered Narnia from the Hundred Years of Winter brought on by the tyrannical magic of the White Witch.
>
> *"We've all seen him."* This indicates what Edmund's faith is based on: his actual experience and the experience of his brother and sisters and fellow Narnians.
>
> *"Lucy sees him most often."* This indicates Lucy's privileged place in Narnian history.
>
> *"And it may be ASLAN'S COUNTRY we're sailing to."* This not only indicates the goal of their voyage but also the focus of their hope and hearts' desires: to live with Aslan.

PROMISE(S) — See OATH(S).

PROVIDENCE — When Cor reflects that ASLAN "seems to be at the back of all the stories" (HHB 14), this is a summary statement of Lewis's theology of providence. How Aslan takes care of all of Narnian HISTORY and the lives of each person and ANIMAL is one of the major themes of the *Chronicles.* Lewis did not believe in luck or fate;[1] rather, he believed in providence: that God's care and foresight are behind all activities, on a scale so large it is impossible for the human mind to comprehend. It seems a misfortune, for example, that SUSAN has lost her HORN somewhere in LANTERN WASTE when the CHILDREN "blunder" out of NARNIA in LWW. But Lewis as subcreator has a providence in mind for events in the *Chronicles,* just as Aslan has a providence for the Narnians. [PC SPOILER] The loss of the horn is actually providential, because in PC it turns up as one of the old Narnian treasures and its note brings the Pevensies back into Narnia to aid CASPIAN against the TELMARINES. In VDT 7, REEPICHEEP comforts EUSTACE with the reflection that his becoming a DRAGON is only a most unusual instance of the way fortune has dealt with royal persons, scholars, and ordinary people in the past: Their fall from prosperity and their recovery lead to the eventual happiness of many. Later (VDT 16) it seems to be fate both that EDMUND, Eustace, and LUCY do not interfere with Reepicheep's departure and also that the children do not follow him but turn to the right and their meeting with the LAMB; both events are more accurately providential. [SC SPOILER] It is indeed providential that RILIAN calls on the three travelers in the name of Aslan to untie him, for this is just the SIGN they have been waiting for (SC 11). In HHB 10, ARAVIS says it is "luck" that the lion did not wound her more seriously, but the HERMIT refers to providence when he says he has "never met any such thing as Luck." And SHASTA shows how well he has learned his lesson when he first attributes his arrival through the mountains into Narnia to luck but corrects himself with, "at least it wasn't luck at all really, it was *Him*" (HHB 12). [LB SPOILER] Finally, it is not tragedy that the SEVEN FRIENDS OF NARNIA and Mr. and Mrs. PEVENSIE are killed in the RAILWAY ACCIDENT, for they are all reunited in the glory of ASLAN'S COUNTRY.

[1]See *The Four Loves,* 126.

PRUNAPRISMIA[1] — CASPIAN's aunt, the red-haired wife of MIRAZ. Caspian dislikes her because she dislikes him. His life is put in jeopardy when she gives birth to Miraz's son (PC 4 and 5).

[1]Readers of Charles Dickens will recognize that Lewis has modeled her name on the characteristic exclamation of Mrs. General in *Little Dorrit:* "prunes and prisms." The suggestion of "prunes" in her name is significant in that Lewis hated prunes. See LITERARY ALLUSION(S).

PUDDLEGLUM — A MARSH-WIGGLE who serves as JILL's and EUSTACE's guide and companion on the journey across ETTINSMOOR in SC. His name implies his marshy origin and his glum outlook (although, ironically, he is considered "flighty" by the other wiggles). A typical Marsh-wiggle, he is all arms and legs, marsh colored, long and thin, beardless, with greenish gray hair that is perhaps more weedlike than hairlike. He wears a high pointed hat with a wide flat brim, and earth-colored clothes. He is, in fact, a skeptical, pessimistic, practical, and down-to-earth sort of person, modeled on Lewis's own gardener, Fred Paxford.[1]

Puddleglum is a Marsh-wiggle of much FAITH, and he trusts to ASLAN's instructions even when things look worst. When RILIAN laughs at Jill's exposition of the SIGNS, Puddleglum jumps to the defense of their mission. His PROFESSION OF FAITH consists of the following: (1) "There are no accidents";[2] and (2) "Our guide is Aslan." When he breaks the witch's spell and she shrieks at him, clearing his head, he is able to assert that though NARNIA and Aslan may not exist, "the made-up things seem a good deal more important than the real ones." This assertion is so strikingly similar to the conclusion of Lewis's essay "The Obstinacy of Belief,"[3] written at about the same time, that one cannot escape the conclusion that Lewis means Puddleglum to be an illustration of the way in which Christians adhere to their faith after it has once been formed.

Puddleglum is also something of a fatalist. Profoundly despondent at Jill's supposed DEATH, he groans that he was meant to be a misfit and as fated to be responsible for Jill's death as he was to eat the Talking Stag at HARFANG.[4] But he accepts his own fault in both matters.

By the end of the ADVENTURE, Jill and Eustace are quite fond of Puddleglum. Puddleglum is among those summoned to the GREAT REUNION in LB.

[1]Lovingly portrayed in *Hooper*, 716–718.

[2]See PROVIDENCE.

[3]In *The World's Last Night*, esp. 29 ff.

[4]See SUICIDE.

PUZZLE — A gray DONKEY, SHIFT's one "FRIEND" and neighbor at LANTERN WASTE, and a main character in LB. Puzzle is a decent sort but has lapsed into the habit of letting the ape do all his thinking for him. He sometimes questions Shift's motives but is easily intimidated and persuaded into doing his bidding. [LB SPOILERS] He shows his innate goodness in several ways: He is concerned about the dead lion from whose skin his coat is made[1] and wants to give it a good burial;[2] he refuses to pretend to be ASLAN, as that would be wrong; and when the EARTHQUAKE and thunderclaps come, he correctly discerns them as SIGNS of divine disapproval. JILL saves him from execution by TIRIAN, and JEWEL the UNICORN speaks to him kindly of HORSEY matters. In ASLAN'S COUNTRY at last he appears as he really is—a lovely gray donkey with an honest face. Aslan, on his final return, seeks out the donkey before all others and whispers something that causes Puzzle to be first humbled and then elated.

[1]Recall Aesop's fable "The Ass in the Lion's Skin."

[2]He says that all lions, even dumb lions, are somewhat solemn because Aslan himself is a lion; thus he shows his basic FAITH in Aslan.

QUEEN OF UNDERLAND — The ruler of the underground kingdom that lies directly beneath NARNIA; she is also known as Queen of the Deep Realm, and the Lady of the Green Kirtle (a loose gown). [SC 4–8 and 16 SPOILERS] The poison green dress she wears foreshadows her transformation into a shiny, great, poisonous green serpent. She is of the same kind as the WHITE WITCH, and one of the "same crew" of EVIL witches. Just as the White Witch intended to rule Narnia with the Hundred Years of Winter, the Queen of Underland intended to rule Narnia through RILIAN as slave-king in his rightful kingdom. The eldest DWARF comments, "[T]hose Northern Witches always mean the same thing, but in every age they have a different plan for getting it."[1] Like her "sisters," she is beautiful and specializes in SWEET seductions. Her realm is an unhappy one populated by enslaved EARTHMEN. She first appears in serpent form, to kill the DAUGHTER OF RAMANDU, Rilian's mother and CASPIAN's wife, and it is as a serpent she dies, at the hands of EUSTACE and the truly disenchanted Rilian. At her death, all of UNDERLAND rejoices.

[1]See WHITE WITCH for a discussion of the recurrence of evil.

QUEST — To quest is to seek; in medieval times KNIGHTS-errant set out in quest of ADVENTURE and KNOWLEDGE. It is important to understand the difference between a quest and an adventure: There may be many adventures within the framework of the quest. JILL's quest for RILIAN is the most complete quest of any in the *Chronicles.* ASLAN charges her with her task and gives her SIGNS for which she must watch. Although she is often tempted to give up her quest, and although she often does not (through her own doubt and inexperience with adventure) recognize the signs when she sees them, at the parliament of owls she recognizes the quest as hers and hers alone and takes it for herself from Eustace, who would have it for his own. At the end of SC, her quest complete, she has a longing for home. Aslan tells her, "You have done the work for which I sent you into Narnia." Since children are brought into Narnia only to save it, in some sense they are all on this same quest. Another word for *quest* is vocation.

R

RABADASH — The crown prince of Calormen during the Golden Age of Narnia, the eldest of the eighteen sons of the Tisroc, would-be suitor of Queen Susan and would-be conqueror of Archenland and Narnia. When first seen in HHB he is a tall, dark, and handsome young man, wearing a "feathered and jeweled turban on his head and an ivory-sheathed scimitar at his side." His face bears an excited expression, and his eyes and teeth have a fierce glint. His name suggests his menacingly wild (*rabid*) and hasty (*dash*) personality and perhaps is a play on *ragabash,* which means "an idle, worthless person." [HHB SPOILERS] Rabadash, in his asinine stubbornness, refuses to accept even the Lion's mercy. For this he is transformed into an ass. Since Rabadash is incapable of acknowledging the Lion, the Lion decides to reach him through the only divine personage the crown prince has any respect for—Tash.[1] Aslan gives Rabadash one more chance by assuring him that he will be healed in the temple of his own god and will remain healed only if he stays within ten miles of the temple. His experience with Aslan at ANVARD changes him into what his people call him to his face—Rabadash the Peacemaker—and what they call him behind his back—Rabadash the Ridiculous. Afterward, to be called "a second Rabadash" in Calormen is to be insulted. (HHB 4, 5, 7, 8, 13, and 14)

[1]See UNIVERSALISM.

RACISM AND ETHNOCENTRISM — C. S. Lewis was a man of his time and socioeconomic class. Like many Englishmen of his era, Lewis was unconsciously but regrettably unsympathetic to things and people Middle Eastern. Thus he sometimes engages in exaggerated stereotyping in contrasting things Narnian and things CALORMENE. He intends this in a broadly comic way, almost vaudevillian. But in our post–September 11, 2001, world, he would, I am sure, want to reconsider this insensitivity.

RAILWAY ACCIDENT — [LB 9, 13, and 16 SPOILER] The SEVEN FRIENDS OF NARNIA and Mr. and Mrs. PEVENSIE are among those killed in LB in a catastrophic railway accident. DIGORY, POLLY, LUCY, EUSTACE, and JILL are on the train—they do not know that the Pevensie parents are in another compartment—and PETER and EDMUND are waiting for them in the station. Lewis seems to have in mind the second-worst rail accident in British history: the October 8, 1952, Harrow and Wealdstone station disaster when 112 people were killed (fourteen on the platform) and 340 hurt when two express trains collided, in northwest London, and a third train ran into the wreckage.[1]

[1]Ascanio Schneider and Armin Mase, *Railway Accidents of Great Britain and Europe* (Devon: David & Charles, 1970), 67–73.

RAM — Surnamed "the Great," he is the son of King Cor and Queen ARAVIS. He becomes the most famous of all the KINGS of ARCHENLAND, though the reason for his fame is not given.[1] His whole STORY is one of the UNFINISHED TALES of NARNIA (HHB 15). "The Ram" is also a name for BACCHUS (PC 11).

[1]According to Lewis's outline of Narnian HISTORY, he succeeds his father and begins his reign in 1050 N.Y. (*The Land of Narnia*, 31).

RAMANDU — A STAR at rest, first referred to as "it." "It was an old man," with silver hair and beard down to the floor. Tall and straight, he is clothed in a ROBE that appears to be made of the fleece of silver sheep, and his feet are bare. He is mild and grave of demeanor, seeming to radiate light and commanding SILENCE and respect. With his DAUGHTER he raises his arms toward the east and sings a ritual song of welcome to the SUN. Afterward a bird flies by and places a bright fruit in the old man's mouth.[1] This is the FIRE-BERRY, which will renew Ramandu each day until he is able to rejoin the great DANCE.

[VDT 13 and 14 SPOILERS] He tells CASPIAN that in order to break the enchantment of the THREE SLEEPERS, it is necessary for someone to sail as far east as possible and to leave there a volunteer who will continue on to the Utter East. He corrects EUSTACE for equating what a star is made of with what it is; and he informs the adventurers that they have already met another star, CORIAKIN, who, he indicates, is earthbound because of some personal failure.

However, he will not reveal the circumstances of Coriakin's fall on the grounds that humans are not meant to know such PRIVATE things about stars. He presses Caspian to decide about sailing east. When the king mentions the weariness of his crew, Ramandu says that great enchantments can be broken only by knowledgeable and willing participants.

[1]Though this image is reminiscent of Isaiah 6:6, the purification of the prophet Isaiah, the dynamic of Ramandu's rejuvenation is very dissimilar.

RAMANDU'S DAUGHTER — See DAUGHTER OF RAMANDU.

RAMANDU'S ISLAND — Also known as RAMANDU'S country, World's End[1] (Island), Island of the Three Sleepers, and Island of the Star (VDT 13–16). This island is the site of ASLAN'S TABLE and the home of Ramandu and his DAUGHTER. Seated at the head of the table, in an enchanted sleep, are the THREE SLEEPERS, lords ARGOZ, REVILIAN, and MAVRAMORN.

[GEOGRAPHY, NARNIAN.]

[1]This island is not to be confused with the true WORLD'S END. The daughter of Ramandu says that though some call Ramandu's Island the World's End, it is more properly the beginning of the end (VDT 11).

RAVEN OF RAVENSCAUR, OLD — Apparently an Old Narnian of some renown. He is in attendance at CASPIAN'S council of war (PC 7). "Ravenscaur" may be a community of ravens, because *scaur* is the Scottish word for "cliff."

REDUCTIONISM, REDUCTIONIST — The idea that things and people can be defined by what they are physically or what they can be used for (a danger in TECHNOLOGY), or that a situation can be reduced to the facts about it. Lewis abhorred the sort of person who "sees all the facts but not the meaning";[1] such persons are always given away by their use of the words "nothing but," "merely," or "only." Lewis even coined a term to describe this process: BULVERISM. One can almost hear the words "nothing but" in EUSTACE's declaration to RAMANDU that a STAR is only a ball of flaming gas; Ramandu speaks for Lewis when he replies that gas is not what a star is, only what it is made of (VDT 14). In her enchantment of the three travelers, the QUEEN OF UNDERLAND tries to reduce ASLAN to "nothing but" a lion, a large CAT, and the SUN into "nothing but" a large lamp (SC 12).

[1]C. S. Lewis, "Transposition," *The Weight of Glory,* penultimate ¶.

REEPICHEEP — A talking mouse who stands between one and two feet tall and has ears nearly as long as a rabbit's. He is the very soul of COURAGE, his head filled always with battles and strategies, HONOR and ADVENTURE. Unlike humans, he has no HOPES or FEARS to contend with. He wears a rapier at his side and, in PC, leads a fully armed band of twelve valiant MICE. He fights bravely at the battle against MIRAZ and the TELMARINES and is brought back more dead than alive. Healed by LUCY's CORDIAL, he is horrified to find that his tail has been cut off. He apologizes to ASLAN for appearing before him in such unseemly fashion, asserting that "a tail is the honor and glory of a Mouse." Aslan restores his tail because of his love for the courage of mice and their ancient kindness in eating away the cords that bound him to the STONE TABLE.

[VDT SPOILERS] By the time of VDT, Reepicheep has been designated Chief Mouse and wears a thin headband of gold with a crimson feather stuck in it. He is one of the heroes of the second battle of Beruna. He reveals that at his birth a DRYAD spoke a prophecy over his cradle that told of SWEET waters, and that he would find what he sought in the Utter East. On the voyage of the *DAWN TREADER* it is Reepicheep who keeps the crew on the path of adventure; he spends his days far forward, gazing always to the east. When EUSTACE insults the mouse by swinging him by the tail, the mouse gives Eustace the first physical punishment of his life, beating him with the flat of his sword. Fair in every way, however, he later upbraids the sailor Rhince for saying "good riddance" to the lost Eustace, and when the DRAGON-Eustace appears he solicits its promise of FRIENDSHIP. His greatest courage comes in his urging the crew to forget their fears and venture into the darkness of DARK ISLAND. When the *Dawn Treader* runs aground at the WORLD'S END, Reepicheep lowers his little boat, throws aside his sword, saying he won't need it anymore, and bids good-bye to EDMUND, Lucy, and Eustace. He vanishes over the wave of the LAST SEA and is never seen again, although Lewis comments that he is probably safe in ASLAN'S COUNTRY.

[LB SPOILER] In LB, he is indeed found to be safe there. Saying "Welcome in the Lion's name," he gives the last summons to "come further up and further in," and Edmund, PETER, and Lucy kneel in greeting as he prevails upon them to come into the GARDEN.

The animals of Narnia spring joyously from the earth, brought forth by Aslan's song of creation. (MN 9)

RESTIMAR — One of the SEVEN NOBLE LORDS. He is judged by the CHILDREN and CORIAKIN to be the beautiful golden statue lying in the lake at the top of DEATHWATER ISLAND. (VDT 2 and 9)

REVELRY — The spirit of revelry is a fundamental part of Narnian life, evoking the medieval world, in which feast days were plentiful. One of the sure SIGNS of enchantment and the work of dark MAGIC is an unending sameness and dreariness, such as can be seen in UNDERLAND and CHARN. It is significant that the chief meeting place in NARNIA is the DANCING LAWN. The complete lack of revelry and the overwhelming sadness of LB is a sign that the very heart has gone out of Narnia.

REVILIAN — One of the SEVEN NOBLE LORDS. He is one of the four Narnian visitors to the Island of the Voices in 2299–2300 N.Y., and one of the THREE SLEEPERS found at ASLAN'S TABLE in an enchanted SLEEP on RAMANDU'S ISLAND. (VDT 2, 11, and 13)

RHINDON — The name of PETER'S sword, one of his gifts from FATHER CHRISTMAS. It has a hilt of gold and fits in a sheath and belt. He uses it to slay FENRIS ULF (Maugrim). (LWW 10 and 12; PC 2)

RHOOP — [VDT SPOILERS] One of the SEVEN NOBLE LORDS, first mentioned by CORIAKIN as one of the four TELMARINE visitors to the land of the DUFFERS in 2299–2300 N.Y. A prisoner on DARK ISLAND, he is first identified by his cry "made inhuman by terror." When he is welcomed aboard the *DAWN TREADER*, his appearance is that of a man who has known harrowing experiences. Pale, gaunt, and wild looking, he wears only a few wet rags and though he is not old his hair is completely white. His wide-open eyes stare "as if in an agony of pure FEAR." He urges those on board to row for their lives, and he is enraged at the naiveté of the sailors who think it would be ideal to live in a place where all one's DREAMS come true. Rhoop tells them that there is a great difference between daydreams and dreams. Ignoring this difference got him imprisoned on Dark Island. Learning CASPIAN's identity, Rhoop begs the favor of never being asked anything about his experiences on Dark Island—a request readily granted.[1]

When the voyagers disembark at RAMANDU'S ISLAND, Rhoop at first stays on board, wanting no more of islands. Later Caspian describes Rhoop's condition to RAMANDU, who suggests that he can give the weary lord what he needs most: a long, uninterrupted SLEEP without a hint of dreams. When Rhoop finally does leave the ship to join the group, he sits down next to Lord ARGOZ at ASLAN'S TABLE. Ramandu gently places his hand on the head of the exhausted lord, and Rhoop slips into a deep and peaceful sleep. He is not mentioned again in the *Chronicles.* (VDT 2, 12, and 14)

[1]In the British and the new HarperCollins editions, Rhoop's request is a simple plea "never to bring me back there," followed by everyone looking in the direction he points and discovering that the Dark Island has disappeared altogether. Because Lewis felt that this might lead children to expect that such fears could be eliminated from life, he rewrote this section for the pre-1994 American editions of VDT, in which Rhoop's request is elaborated, Caspian responds, and the island recedes in the distance. For a complete discussion of these differences, see DREAM(S).

RIGHT AND WRONG — Lewis believed there was a clear distinction between right and wrong; between morality and immorality; and between good acts and bad acts. With the proper education, any person should be able to develop the STOCK RESPONSES to allow that person to know that good is to be done and EVIL is to be avoided. The instinctive and educated knowledge of what is good and what is evil is called conscience. In the *Chronicles,* it is often expressed as the feeling "deep down inside" that something ought to be done or ought to be avoided. [MN SPOILERS] When DIGORY asks his Uncle ANDREW whether anything was "wrong" about Mrs. LEFAY, Andrew replies with a chuckle, "Well, it depends on what you call *wrong.*" The evil in Andrew's outlook and activities is echoed and amplified to its most terrible consequences in Jadis's use of the DEPLORABLE WORD to destroy CHARN. Like him, she has broken OATHS and paid a terrible price to learn her MAGIC; like Andrew with his guinea pigs, she does what she wishes with her people because she thinks she owns them; like him, she pleads that her role in society exempts her from ordinary moral considerations; and as with him, her greedy look betrays her lack of principle. Digory is capable of following either the right or the wrong path; although he is initially overcome by the witch's beauty, she eventually repels him. Ultimately, he is tested in the GARDEN of the west, from which he must bring a silver apple back to ASLAN without eating or stealing it. When Digory SMELLS the silver apple he has plucked in the garden of the west, he has "a terrible thirst and hunger . . . and LONGING to taste the fruit." It suddenly seems right that he should do so, even though he knows it to be wrong. Digory has been brought up correctly, however, and knows it is wrong to steal and wrong to break promises. He resists, but Jadis does not. When Digory returns the apple, untouched, to Aslan, the Lion explains that the fragrance of the TREE OF PROTECTION will be a horror to the witch, because "that is what happens to those who pluck and eat fruits at the wrong time and in the wrong way. The fruit is good, but they loathe it ever after." In Letter IX of *The Screwtape Letters,* Lewis writes this lament for the tempter Screwtape: "All we can do is to encourage the humans to take the pleasures which [God] has produced, at times, or in ways, or in degrees, which He has forbidden."

The problem of right and wrong does not exist in ASLAN'S COUNTRY, because just as there is no evil in that place, it is also impossible to do or want anything wrong there (SC 16 and LB 13).

RILIAN — A KING of NARNIA, the son of King CASPIAN X and the DAUGHTER OF RAMANDU, and known as King Rilian the Disenchanted. Throughout most of SC he is held captive by the enchantment of the QUEEN OF UNDERLAND, who killed his mother when the prince was about twenty years old. Seeking to avenge his mother's DEATH, Rilian instead falls under the witch's spell and goes to live with her for ten years in her underground kingdom. Rilian is among those summoned to the GREAT REUNION in LB.

RING(S) — Four paired rings—two yellow, two green—that ANDREW Ketterley has made from the MAGIC dust in Mrs. LEFAY'S box and that enable DIGORY and POLLY to make their TRANSITION to the WOOD BETWEEN THE WORLDS

(MN 2, 3–8, 9, 13, and 15). Shiny and beautiful, they work only when they touch bare skin (although, as with electricity, the magic will run through one person to another if they are touching). Uncle Andrew, a typical magician, has a limited understanding of how the rings work—he thinks that the yellow rings lead out of our world, and the green rings lead back. Actually, they are far less specific. Since the rings are made of material that originated in the Wood between the Worlds, the yellow rings want to get back to the wood, while the green rings want to go out of the wood. In a telling statement about the essential triviality and trickery of any magic in the presence of God, ASLAN tells the CHILDREN that they "need no rings" when he is with them and commands them to bury the rings, which they do—at the foot of the apple TREE in Digory's backyard. Although the SEVEN FRIENDS OF NARNIA resolve to use the rings again and recover them, they are never used (LB 5).

RISHDA TARKAAN — The CALORMENE captain who joins with SHIFT in the plot to enslave NARNIA by means of the lie about the false ASLAN. He is fawning and thoroughly corrupt; his first actions in LB are to steal TIRIAN's crown and then to tell the lie that he has captured Tirian and JEWEL by skill and COURAGE. He does not really believe in either TASH or Aslan (see TASHLAN) and says that Aslan means no more and no less than Tash. He calls the DWARFS "CHILDREN of mud" and offers them as a burnt sacrifice to Tash. He is horrified by the sight of the real Tash, whose name he had invoked without FAITH, and disappears under the arm of the GOD to Tash's "own place," presumably HELL. (LB 3, 9, 12, and 15)

RIVER-GOD — He arises from the waters of the GREAT RIVER (accompanied by his daughters, the NAIADS) at ASLAN's command to the waters of NARNIA to be divine, and he is one of the seven beings invited by Aslan to the first council (MN 10). He has a deep voice, a weedy beard, and a larger-than-human head crowned by rushes. He considers the bridge at Beruna to be his chains, and at Aslan's command BACCHUS destroys it. (PC 9 and 14)

ROBE(S), ROYAL — Narnian clothes in general look, feel, and SMELL good (there is no scratchy wool flannel, starch, or elastic in all of NARNIA [LB 12]; Lewis is making an AUTOBIOGRAPHICAL ALLUSION to clothes he hated to wear when he was a child). Royal robes are no exception. At their coronation King FRANK and Queen HELEN are dressed in "strange and beautiful" clothes, and their long trains are attended by DWARFS and NYMPHS (MN 14). The clothes of the Narnian embassy to TASHBAAN are comfortable and colored in strong earth tones (HHB 4). In comparison to the fabulous Middle Eastern–style clothes of the CALORMENES and the expensive, jewel-encrusted clothes of the enchanted royalty at CHARN (MN 4), the royal robes of Narnia are simple and direct. In fact, it may be said that in Narnia "the man makes the clothes" rather than the other way around: King LUNE greets ARAVIS in his oldest clothes, and the young Tarkheena (who is used to the formality of Calormen) is a bit put off (HHB 15). In comparison to English clothes, Narnian robes are brighter and easier to wear (LB 12). In a preview of their later elevation to royalty, the four Pevensie CHILDREN borrow coats from the WARDROBE to wear into Narnia. Too big for them, the coats hang down to their heels and look like royal robes. Each thinks the other looks much better and more suited to the landscape (LWW 6). At the end of PC (15), they must remove their royal robes and change back into their school clothes in order to reenter ENGLAND from Narnia. None are at all happy about this prospect, and a few TELMARINES typically make fun of their drab outfits. Lewis seems to have the same feeling about school clothes as he does about SCHOOLS; the little girls at GWENDOLEN'S school in Beruna wear tight, itchy uniforms. TIRIAN thinks JILL'S and EUSTACE'S clothes are odd and dingy (LB 4 and 5), and in SC 3 the children themselves look dingy in comparison to their Narnian surroundings.

[BANNER, STANDARD, CROWN, CORONET.]

ROGIN — See DUFFLE.

ROONWIT — A great, golden-bearded CENTAUR whose voice is as deep as a bull's. Like all Centaurs, he can read the STARS; indeed, in Old Norse, *roonwit* means "he who knows how to read the sacred language." He tells TIRIAN that he has seen bad signs in the stars all year, and there are disastrous alignments of planets in the heavens. He drinks to ASLAN and truth, and because the stars never lie, he concludes that the rumor that Aslan is in the country is untrue. Roonwit is sent to rally Narnian forces against the CALORMENES, and much depends on him. But the Calormenes take CAIR PARAVEL and a Calormene arrow kills Roonwit. FARSIGHT the eagle shares the Centaur's last hour and bears his last message to Tirian: "Remember that all worlds draw to an end and that noble DEATH is a treasure which no one is too poor to buy." He is among those who pass the JUDGMENT into ASLAN'S COUNTRY and is the first to raise the cry to go "further in." (LB 2, 6, 8, and 14)

RUMBLEBUFFIN — A good GIANT (most likely a member of the Buffin family), freed by the BREATH of ASLAN from imprisonment as a statue in the WHITE WITCH'S courtyard. Aslan enlists him to break down the gate and the towers of the witch's castle. In true giant fashion, he clubs and tramples in battle, but always on the side of good. He is rewarded and honored at the feast of the crowning. (LWW 16 and 17)

RYNELF — A sailor on the *DAWN TREADER*, fashioned by Lewis after the dutiful, articulate, and experienced seaman of many a sea STORY. (VDT 1, 12, 13, 14, 15, and 16)

S

SALAMANDERS OF BISM — Small, DRAGON-like creatures that live in the rivers of fire in Bism (SC 14) and, like the SUN, are too white-hot to be looked at directly. They are very eloquent. When a hissing voice announces the closing of the rift of Bism, Lewis notes that the four travelers are not sure if it is the voice of Fire itself or of a salamander. Lewis reflects an ancient belief that there is a kind of creature that inhabits each of the four elements: Salamanders or Vulcans in fire, Gnomes or DWARFS in earth, Sylphs or SILVANS in air, and Undines or NYMPHS in water.

SALLOWPAD — The large raven in the Narnian embassy to CALORMEN in 1014 N.Y. (HHB 4 and 5). His is the voice of wise counsel in the debates of EDMUND, SUSAN, and the Narnians who are hostage to RABADASH's whim. His name means "yellow footed."

SATYR(S) — In Greek and Roman MYTHOLOGY, reddish brown wood GODS who have goat legs, pointed goat ears, and sprouting horns and who are considered to be quite lascivious; they look more like goats than do FAUNS, who are only half-goats. One of the NINE CLASSES OF NARNIAN CREATURES, they are among the wild people who emerge from the wood at ASLAN's command to NARNIA to awaken (MN 10).[1]

[1]Like all mythological creatures in the *Chronicles,* they are not expressly created by Aslan. See CREATION OF NARNIA.

SCEPTRE — The royal staff or rod, the symbol of the authority of the EMPEROR-BEYOND-THE-SEA, ASLAN's Father. The words of the Deep MAGIC are engraved on it (LWW 13). Nothing more is said of it, but as a symbol of good power, it is to be contrasted with the WHITE WITCH's WAND.

SCHOOL(S) — Lewis's own experiences in school were not happy. He learned French and Latin from his mother, and other primary subjects from a governess, all at home. A month after his mother's death in late August 1908, the nine-year-old Lewis began almost six years of misery in boarding schools. So bitter was his experience of his first headmaster, the Rev. Robert Capron (who was certified insane in 1910 and died in 1911), that Lewis was able to forgive this cruel ADULT only in the last years of his own life. In the fall of 1914 Lewis began three of the happiest years of his life living with his tutor, W. T. Kirkpatrick.

So Lewis considered school a necessary EVIL, and the schools mentioned in the *Chronicles* reflect this. The trouble with schools in general, Lewis hints, is that instead of teaching values through STORIES they teach dry subjects from ugly books. When Professor DIGORY KIRKE wonders in LWW 17 what they teach CHILDREN these days, he is implying that they certainly do not teach them about ADVENTURE, and especially about the adventure of FAITH. Children in the Narniad simply do not want to go to school. In PC, the Pevensie children consider themselves lucky to have been swept into NARNIA just at TERM-TIME, and SHASTA (now Cor) laments that BREE will have a much better time because he does not have to get an education (HHB 14). MYTHOLOGICAL creatures seem to be hieroglyphs[1] for the free spirits of children in at least two instances: One of the first commands of the new KINGS and queens is that the young DWARFS and SATYRS be released from the bondage of school; when GWENDOLEN leaves the girls' school at Beruna to join ASLAN, the MAENADS free her from her tight school uniform.

Although Lewis dislikes most schools, he particularly dislikes modern schools such as EXPERIMENT HOUSE. Here bullies and GANGS are treated like interesting psychological cases instead of being punished, and ordinary children live in terror (SC 1, 3, and 15). Although JILL has somehow managed to see pictures of mythological creatures, EUSTACE is a victim of his education, at least at first. He actually enjoys looking at pictures of foreign children doing exercises and knows nothing of adventure or HONOR. In fact, he is well on his way to becoming a model civil servant in the government of ENGLAND when he is transported into Narnia. It is here that his real education begins, and Jill notes that he seems more a product of Narnia than of Experiment House.

TRUMPKIN is compared to a crusty but lovable old teacher (SC 4). Doctor CORNELIUS is the only likable educator in the *Chronicles.* Although he teaches CASPIAN the usual dry subjects, he also teaches him the truth and awakens him to his VOCATION.

[AUTOBIOGRAPHICAL ALLUSION(S).]

[1]For more on Lewis's use of hieroglyphs (picture writing), see TALKING BEAST(S).

SEA PEOPLE — Inhabitants of the LAST SEA (VDT 15 and 16), they are a race of aquatic beings (not to be confused with the MERMEN and mermaids, who are amphibious). Their cities are sea-mounds, their farms and pastures the shallows, their places of danger (and therefore of ADVENTURE) the sea valleys. LUCY is the first to see Sea People, a HUNTING party comprising royalty, KNIGHTS and ladies, all mounted on seahorses and "hawking" with hunting fish instead of falcons. They appear to be a fierce people, immediately challenging the threat of the *DAWN TREADER* sailing overhead. It is in answer to this challenge that the courageous REEPICHEEP dives overboard and thereby discovers that the waters of the last sea are SWEET. Superstitious (as are most men of the sea), Lord DRINIAN thinks the Sea People unlucky. As if to contradict this suspicion, Lewis seems to make a great deal of the instant FRIENDSHIP that grows up between Lucy and the Sea Girl, a fish-herdess wandering alone with her flock in the shallows of the last sea. Lucy is ever the figure of the person who sees beyond what other people see to the possibilities of love and understanding.

SEA SERPENT — A MYTHOLOGICAL sea monster that appears horrifying but acts stupidly (VDT 8). Encountered by the *DAWN TREADER* on its journey from Burnt Island to DEATHWATER ISLAND, it is attacked by EUSTACE and pushed off the boat by the whole crew. The Sea Serpent is mentioned with the squid and the KRAKEN as the three dangerous beasts of the deep that are feared by the SEA PEOPLE.

SECRET HILL — In the heated exchange between the WHITE WITCH and ASLAN (LWW 13), the witch alludes to the fact that the words of the Deep MAGIC are deeply carved, like runes, into the fire-stones of the Secret Hill. No specific place is probably intended here, but Lewis may have used the term to evoke in the IMAGINATIONS of his readers pictures of annual pagan rites throughout the British Isles in which the old year's fires were extinguished and the new fire was kindled at a sacred place, usually a low, round hill. The fire-stones were set in a permanent ring on the crown of the hill. No BIBLICAL ALLUSION is intended.

[WORLD ASH TREE.]

Sorely tempted to disobey Aslan and eat a silver apple from the Tree of Protection, Digory wrestles with his conscience. He is watched by the evil witch and a wonderful, mysterious bird. (MN 13)

SEVEN BROTHERS OF THE SHUDDERING WOOD — Red DWARFS who live among rocks and fir TREES on the northern slopes of the mountains that border NARNIA on the south (PC 6 and 7). They are blacksmiths and armorers and work at an underground forge, where the blows cause the surrounding woods to shudder. Once they become convinced of CASPIAN's kingship, they give generous gifts of armor of the finest workmanship. They are also in attendance at the great council and take active part in the WAR OF DELIVERANCE.

SEVEN FRIENDS OF NARNIA — They are DIGORY Kirke, POLLY Plummer, PETER Pevensie, LUCY Pevensie, EDMUND Pevensie, EUSTACE Clarence Scrubb, and JILL Pole.[1] [LB 4, 12, and 16; SPOILER] These seven SONS OF ADAM and Daughters of Eve are killed in a RAILWAY ACCIDENT in 1949. They are first called friends of NARNIA by TIRIAN, who calls for their help in saving Narnia from its last great challenge. Tirian is somehow able to see through to our world and, as if in a DREAM, glimpses the seven friends. They in turn recognize Tirian as a Narnian, and Eustace and Jill are dispatched to his aid.

[1]See SUSAN PEVENSIE for a discussion of her notable exception from this group [LB SPOILER].

SEVEN ISLES — A cluster of islands, and the third port of call of the *DAWN TREADER* (VDT 2). First mentioned in LUCY's daydream about past voyages in the GOLDEN AGE OF NARNIA (PC 8), they are the furthest limit of the special visionary powers of the HERMIT OF THE SOUTHERN MARCHES (HHB 13). The two western islands are given names: Muil, the westernmost; and Brenn, the capital city of which is Redhaven. The other five are not named. When the elderly King CASPIAN IX sets out on his last voyage in order to seek ASLAN's counsel, he intends to visit the Seven Isles (SC 3).

[GEOGRAPHY, NARNIAN]

SEVEN NOBLE LORDS — TELMARINE lords, feared by MIRAZ because of their friendship with CASPIAN IX (PC 5). They are REVILIAN, ARGOZ, and MAVRAMORN (the THREE SLEEPERS); OCTESIAN, who met his end on DRAGON ISLAND; RESTIMAR, transformed into a golden statue on DEATHWATER ISLAND; RHOOP, the victim of DARK ISLAND; and BERN, who is made duke of the Lone Islands. Together they are sent by Miraz (who hopes to be rid of them) to explore for any lands that might exist beyond the Eastern Sea, the end of the then-known world. They are the only Telmarine lords who are not afraid of the sea, but like all Telmarines they were forbidden by Miraz to learn any nautical skills. Thus they must buy a ship from GALMA and man it with Galmian sailors before they can begin their voyage. They have not returned by the time Caspian is told about them by Doctor CORNELIUS, on the night that PRUNAPRISMIA has a son. Caspian takes a coronation day OATH to seek for them and avenge their DEATHS with HONOR; the story of VDT is the record of how he fulfills this QUEST (VDT 2).

SEXISM, SEXIST — The *Chronicles* have a number of superficial remarks against persons because of their gender (usually against women), but Lewis's insights into character often reveal a basic sympathy for the equality of women. This sympathy is especially apparent in the last four books

(SC, HHB, MN, and LB), in which the female characters are much more realistically drawn than in the earlier books. It is important to note that although HHB was published after SC, it was actually written first; thus ARAVIS is the pivotal female character in Lewis's growing insight into feminine character.

In the first three books, LUCY especially shows COURAGE, but to Lewis's mind (and as CORIN remarks in HHB) this is still a boy's qualities, not a girl's. In LWW, SUSAN and Lucy are not to be in the battle, not because women aren't brave enough, but because "battles are ugly when women fight" (LWW 10). Mrs. BEAVER comments that it's "just like men" (LWW 10) to be discussing swords when the tea is growing cold. In PC, EDMUND says to TRUMPKIN and PETER, "That's the worst of girls. They never can carry a map in their heads." Lucy retorts, "That's because our heads have something inside them" (PC 9). When King Peter kisses TRUFFLEHUNTER, Lewis quickly explains that this is "not a girlish thing," because such COURTESY is expected in NARNIA.

With the introduction of Eustace in VDT, the subject of sexism becomes even more complex. When CASPIAN gives Lucy his own fine cabin because she is a girl, EUSTACE complains that to do such a thing is really demeaning to girls. Of course his real motivation is selfish—he wants the fine cabin for himself. Lucy, for her part, has her own sexist opinion of men. "That's the worst of doing anything with boys," she says. "You're all such swaggering, bullying idiots" (VDT 8). When the DUFFERS threaten Lucy in order to get her to undo the spell of invisibility, she is the only one who behaves bravely. Her innate good sense cuts through to the practical heart of the matter, and the boys actually seem to be embarrassed at her courage and their COWARDICE.

With the introduction of Aravis, Lewis seems to have undergone a real change in his experiences of and appreciation for women. In HHB, she flees from CALORMEN, where women are property, to seek freedom in Narnia, where women are equal to men. She refuses to be pushed around (nor will her HORSE, HWIN). To SHASTA's "Why, it's only a girl!" she bristles, "And what business is it of yours if I am *only* a girl?" (HHB 2). She is "proud . . . hard . . . and true as steel" (HHB 6), and she never once loses her head. In contrast to LASARALEEN, the stereotypical fluttery female interested in clothes and men, Aravis has always been more interested in bows and arrows and horses and DOGS and swimming, considered boys' things by Lewis's generation. It is also in HHB that Lewis puts sexist remarks in the mouths of fools: For instance, RABADASH shows his complete misunderstanding of Aravis herself and women in general when he says "it is well known that women are as changeable as weathercocks" (HHB 8).

There is a subplot in HHB between the horses Hwin and BREE, in which Bree's condescension toward the mare shows him up as full of pride and VANITY. Corin, something of a sexist himself, gives implicit though patronizing praise for Lucy when he says that Lucy is "as good as a man, or at any rate as good as a boy. Queen Susan is more like a grown-up lady" (HHB 13).[1] Queen Lucy fights in the Battle of ANVARD from among her fellow archers. When Lucy and Aravis meet, they like each other instantly and go off to talk about "all the sort of things girls do talk about on such an occasion" (HHB 15), which Lewis assumes to be clothes. This is a lapse in Lewis's newly nonsexist attitude. It is more likely that the girls would compare notes about battles and journeys, since that is what they have been involved in.

In SC, JILL, the primary female character, is perhaps the most real of all the girls in the *Chronicles,* and her actions are both brave and fearful. Eustace has improved but not totally reformed, and he echoes Edmund's comment in PC about maps when he says, "It's an extraordinary thing about girls that they never know the point of the compass." This may be intended as satire on Lewis's part, since Eustace doesn't know which direction is east any more than Jill does. Lewis adds that Jill indeed does not know her compass points, but that this does not necessarily apply to all girls. He shows here a nice sensitivity toward his girl readers that is missing from the earlier books. In contrast to Lucy and Susan, who fought with bows and arrows only, Jill carries a knife and is glad of it. Indeed, she becomes more and more assertive as the STORY progresses.

[SPOILER] In a scene disappointing to many girl readers, Jill does not slay the QUEEN OF UNDERLAND but remains on the sidelines trying not to faint or cry. Although this is certainly a very real reaction, Eustace, PUDDLEGLUM, and RILIAN are able to overcome their own distaste and do the deed. The three males leave the witch's castle with drawn swords, and Jill with drawn knife; even in weapons Jill is creeping up to them in equality since Lucy's day, but she's not there yet. At the beating of the GANG, Jill still has a unique weapon—a switch–turned–riding crop—while the boys use the flats of their swords.

POLLY, in MN, is not the conventional turn-of-the-century

girl. She has made a cave for herself with packing cases, is independently exploring her attic, and keeps treasures that include "a story she was writing" (MN 1). Uncle ANDREW, the EVIL* MAGICIAN, is a blatant sexist: He says that morality is for little boys and servants and women and people in general. Later, when DIGORY's defense of basic RIGHT AND WRONG gets through to his uncle, Andrew recovers with a perfect BULVERISM: You only say that because you were brought up among women and learned this natural morality from old wives' tales.

By the time of LB, Eustace has gained more respect for Jill's abilities. She succeeds in killing and preparing a rabbit, the first instance of HUNTING by a woman in the *Chronicles.* She has become self-sufficient and has learned her compass points. It is she who keeps the travelers on the path, and TIRIAN puts her in the lead as the best guide of them all. She is now brave enough to investigate the STABLE on her own, with knife drawn and ready. She is compassionate and supportive, saving PUZZLE from DEATH at the hands of Tirian. When Tirian tells Jill and Eustace that they must go home to avoid the bloody battle that will follow, Jill protests that they have not yet done what they came to do, and she does not shrink from danger. Although both children later admit that they are scared, Jill declares that she would "rather be killed fighting for Narnia than grow old and stupid at home"[2] (LB 9).

[1]This is also a comment on the difference between Lewis's conception of the values of ADULTS and CHILDREN, and [SPOILER] a foreshadowing of Susan's later fall from status as one of the SEVEN FRIENDS OF NARNIA.

[2]See ADULT(S) and AGING AND DISABILITY.

SHASTA — [SPOILERS] The CALORMENE name for Prince Cor, the lost son of King LUNE OF ARCHENLAND, elder twin of CORIN THUNDER-FIST, unwilling companion, then friend, and finally husband of ARAVIS Tarkheena, and father of RAM the Great. His recovery of his identity as free-born son of a northern king and his preparation to become the compassionate and just ruler of Archenland is the main STORY in HHB. His becoming a horseman is a picture of how much he has to learn to control himself and even push himself in order to gain true freedom. BREE, as an experienced HORSE, can teach him much in this regard, but Bree's VANITY and his reactions to his long years of slavery in Calormen aren't enough to form Shasta as an adult. Before he is ready to become a king, the boy needs to meet Aravis and HWIN, to pass through the FEAR of the tombs and the trial of the great desert, and finally to be sent on by the HERMIT to experience ASLAN both in the foggy night on the pass (where he has the most important NUMINOUS experience of the Narnian "TRINITY" in the entire *Chronicles*) and in the bright morning of southern NARNIA. Like Bree, Shasta has picked up bad habits in Calormen and he knows little about the ways of free people; his harsh upbringing leads him to expect nothing but ill treatment and duplicity from ADULTS. But he does have an innate sense of faithfulness and nobility, the foundation on which his character will be built. And though he has many lessons to learn about court manners and COURTESY and about accepting his VOCATION to be king and the like, he has learned the great lesson of his life, the lesson of OBEDIENCE, of PROVIDENCE, of HOPE, and of HISTORY: Aslan "seems to be at the back of all the stories" (HHB 14).

SHERLOCK HOLMES — The famous fictional detective created by Sir Arthur Conan Doyle. He is mentioned by Lewis in the second paragraph of MN 1 to set the time of the story. Since the *Chronicles of Narnia* is a work of fiction, NARNIA really exists in the world of fiction in which Holmes is also real.

[BASTABLES.]

SHIFT — [LB SPOILERS] An ancient, clever, ugly ape who lives by CALDRON POOL in an otherwise uninhabited area of LANTERN WASTE. The name "Shift" is indicative of his manipulative personality: He is "shifty"—underhanded, sneaky, and a liar; and he has a great facility for "shifting" meaning—he redefines the meaning of freedom and ASLAN to suit his purposes and he wickedly merges Aslan and TASH into TASHLAN. He is completely selfish and totally corrupt, and he degenerates perceivably and rapidly from his first conception of PUZZLE as a false Aslan to his DEATH at the hands of Tash. His relationship with Puzzle, which he calls a FRIENDSHIP, is really that of master and servant. Shift does not believe in anyone or anything but himself. As an ape, he looks like a man but isn't; in fact, he tries to pass himself off to the ANIMALS as a wise old man. At the STABLE, he is dressed garishly, wears a paper crown, and calls himself "Lord Shift, the mouthpiece of Aslan." Although never as much in control of the deception as he thinks he is (Shift makes several "Freudian" slips along the lines of "I want—I mean, Aslan wants . . ."), as LB progresses he becomes an alcoholic and yields some power to RISHDA. By the time he gets to the evening assembly at the stable, he has a hangover and it is no problem for Tirian to toss the ape through the door to his death.

[Wine.]

SHIP, THE — See Leopard, the.

SIGN(S) — Aslan gives Jill and Eustace four signs[1] by which he will guide them in their quest for the lost Prince Rilian: (1) Eustace will immediately meet an old, dear friend, whom he is to greet at once; (2) the two must go north until they come to the ruined city of the ancient giants; (3) they will find there written instructions that they must carry out; and (4) they will know Rilian because he will be the first person to ask them to do anything in Aslan's name. Jill is instructed to remember the signs and believe them, because they will not be anything like what she expects them to be. She must repeat them day and night so that she will not forget them in the thick (in comparison to Aslan's country) air of Narnia. [SPOILERS] She promises, but once she's in Narnia, almost everything gets in the way of remembering the signs and she is continually tempted to give it all up. The comfort and excitement of the trip to Narnia on Aslan's breath distract her from the first sign; they fail immediately to recognize the ruined city even when they are walking across it; they do walk right through the words "UNDER ME" but think they are deep trenches; when they do recognize the fourth sign, they cannot believe they are meant to obey the madman giving it. It is only due to the insistence of Puddleglum that any of the signs are remembered.

[Earthquake(s); Memory.]

[1]Lewis virtually quotes Deuteronomy 6:4–8, and it is almost impossible to escape the connection between Jill's signs and the Ten Commandments of the Judaeo-Christian tradition. See BIBLICAL ALLUSION(S).

SILENCE, SILENT — Lewis frequently uses silence as a way to describe the solemnity of a NUMINOUS person or activity or as a tool by which characters may face themselves. Silence is often the best answer to an overwhelmingly good or bad situation, one that no words can describe. There is deep silence between CASPIAN and Doctor CORNELIUS at the revelation of the true HISTORY of NARNIA (PC 4). After ASLAN tells LUCY she will find him bigger the older she grows, she is so happy she does not want to speak (PC 10). In HHB 11, SHASTA's experience of the glorified Lion causes him to fall at its feet. "He couldn't say anything but then he didn't want to say anything, and he knew he needn't say anything." In MN 3, the WOOD BETWEEN THE WORLDS is "the quietest wood you could possibly imagine." CHARN, on the other hand, is filled with a "dead, cold, empty" silence (4). At the CREATION OF NARNIA, FRANK has frequently to call for silence to listen to the creation song: "Watchin' and listenin's the thing at present; not talking" (9). At the selection of the pairs of TALKING BEASTS from the DUMB BEASTS, there is complete silence for the first time since the creation began (10).

SILENUS — In Greek and Roman MYTHOLOGY, the drunken attendant of BACCHUS (the GOD of REVELRY). He rides a DONKEY. He is an expert musician and a regular summer visitor to NARNIA and is, of course, absent during the Hundred Years of Winter. In PC 11 and 14, he shows up with Bacchus at the third DANCE. Old and enormously fat, he continually falls off his donkey and calls for refreshments—which seems to cause grapevines to grow luxuriantly and produce WINE immediately. Lucy sees a book titled *The Life and Letters of Silenus* on TUMNUS's bookshelf (LWW 2).

SILVANS — In Greek and Roman MYTHOLOGY, the TREE-PEOPLE, the spirits of trees (PC 14). They are created (along with the DRYADS, HAMADRYADS, WOOD PEOPLE, and GODS and goddesses of the wood) by ASLAN's command to NARNIA to awake (MN 10).

SILVER CHAIR — In SC 11, RILIAN is tied each night to a "curious silver chair" in his apartments in UNDERLAND. His first free act is to destroy the MAGIC chair, and as it breaks under his sword "there comes from it a bright flash, a sound like small thunder, and . . . a loathsome smell."

SILVER SEA — The very last part of the LAST SEA (VDT 16). It is named for the white lilylike flowers whose indescribably fragrant SMELL, all-pervasive but not overpowering, fills the crew of the *DAWN TREADER* with energy. Henceforth no one needs to eat or SLEEP. Another name by which this region was initially known is Lily Lake.

SILVER TREE — A TREE made of silver that grows from three silver half-crowns[1] (about the size of a U.S. silver dollar) that fall from ANDREW Ketterley's pocket into the rich Narnian soil (MN 11 and 14). Along with the GOLDEN TREE, it grows inside the cage where Uncle Andrew is kept.

[1]A sixpence also falls from Uncle Andrew's pockets. Since no other trees are mentioned, it can be assumed that this silver coin was also one of the seeds of the silver tree; it is about the size of a U.S. nickel.

SLAVE TRADE — The appearance of the slave trade in NARNIA always signals decay in the social structure. Lewis comments that the worst part of being a slave is that one loses one's own willpower. In VDT, there is a thriving slave market at Narrowhaven, in which CHILDREN and ADULTS are sold as impersonally numbered lots. Governor GUMPAS, who excuses the practice as an "economic necessity," is removed from office by CASPIAN and replaced with Lord BERN, who abolishes slavery in the Lone Islands. In HHB, the TARKAAN and ARSHEESH haggle over the selling price of SHASTA, which implies that slavery is a common practice in CALORMEN. Calormene slaves are treated very badly, as evidenced by the fact that RABADASH would think nothing of hanging an idle slave. Slavery is widespread in LB under the rule of the degenerate Calormenes, and even TALKING BEASTS are harnessed for work.

SLEEP — Sleep (and its counterpart, waking) is a complex subject in the *Chronicles* and operates on several different levels of meaning. In SC 4, Lewis offers a PRACTICAL NOTE when he says, "The sleepier you are, the longer you take about getting to bed." And he offers a wonderfully realistic description of SHASTA drifting off to sleep in the desert (HHB 6). Sleep may also be a gift of oblivion, such as the sleep given Lord RHOOP after his years of nightmare on the DARK ISLAND (VDT 14) and the sleep given to Uncle

Andrew (the only gift he is able to receive), which is a temporary rest "from all the torments [he] has desired for [himself]" (MN 14). For the three sleepers, sleep is an enchantment brought on by their disrespectful handling of the stone knife (VDT 13). The apple from the Tree of Protection gives Mabel Kirke her first "real, natural, gentle" drug-free sleep (MN 15).

Beyond these meanings, however, there is the hint that what we think of as wakefulness is only a dream, and that there is another state in which we may be truly awake.[1] When Lucy first hears Aslan's name in LWW 7, she has "a beginning of vacation, waking up in the morning feeling."[2] There is a dreamlike quality to her first vision in PC 9, in which the trees almost talk. In her second vision (PC 10), she knows that the trees are not quite awake, but that she herself is "wider awake than anyone usually is." When Polly and Digory see the swirling vision of Aslan's glory in MN 15, they suddenly feel as though they have never "really been . . . alive or awake before." That this waking has to do with entrance into Aslan's country is powerfully apparent in LB 14, when Jill and Eustace remember that Father Time, whom they met as he lay dreaming in Underland, is to wake on the day the world ends. Aslan adds mysteriously, "While he lay dreaming his name was Time. Now that he is awake he will have a new one." This new name is never revealed, but the implication is that time itself is only a worldly dream. And if there was ever any doubt, at the end of LB Aslan fulfills Lucy's original feeling back in LWW and tells the inhabitants of his country that now "the term is over: the holidays have begun. The dream is ended: this is the morning" (16).

[1]See Plato.

[2]This is the first of many parallels Lewis makes between term-time and sleep, and vacation time and waking. Lewis disliked schools, and he compared life on earth to a finite series of lessons, and death and entrance into heaven to the beginning of holidays and joy.

SMELL(S) — From the smell of fresh fish frying for the Beavers' lunch to the delicious smell of the garden of the west, the smells of Narnia pervade the *Chronicles.* Smells often communicate visions and emotions: Lucy describes the smell that wafts in from the hills of Ramandu's Island as "a dim, purple kind of smell"; and later, "the smell of the fruit and the wine blew towards them like a promise of all happiness." The lilies of the Silver Sea smell wild and lonely, and the sea itself smells tingling and exciting. A breeze from Aslan's country bears both a smell and a music that would break your heart if it could be spoken, but it is not sad. Aslan himself—the transformed wooden horse in Jill's dream—fills the room with "a smell of all sweet-smelling things there are." Along with all the wonderful smells, there are terrible smells: the "loathsome smell" of the disintegrating silver chair; the sickeningly sweet smell of the Queen of Underland's incense; the smells of unwashed dogs and piles of garbage in the streets of Tashbaan; and the smell of dead flesh that accompanies the approach of Tash. The ordinary smells are wonderfully familiar: the "nice, honest, homely smell" of the stable; the fusty smell of the meeting room of the parliament of owls; the animal smell of Aslan, who invites Bree to sniff him in friendship. Finally, there is the powerful and mystical smell of the Tree of Protection, whose fragrance is "JOY and life and health" to Narnians, but "DEATH and horror and despair" to the WHITE WITCH.

[DOMESTICITY; LONGING; SOUND(S); TACTILE IMAGE(S).]

SMOKING — Lewis makes several references to pipe smoking in the *Chronicles.* TRUMPKIN enjoys smoking his pipe, and its tobacco is "fragrant," which implies Lewis's approval of the activity. The Red DWARF brothers DUFFLE and Bricklethumb also take a pipe after dinner, and Poggin (after first asking JILL's and TIRIAN's permission) lights an after-breakfast pipe. Smoking is not limited to Dwarfs; MARSH-WIGGLES smoke a dark blend (possibly mixed with mud) that is so heavy the smoke floats down instead of up. Master Rhince, who worries that his tobacco won't last the trip, is the only human being in the *Chronicles* to smoke a pipe. Lewis was a smoker from his YOUTH; the effects of smoking led to most of his health problems and to his death.

SON OF ADAM, DAUGHTER OF EVE — Used interchangeably by Lewis with the word *human* to identify human boys and girls. Lewis believed that man and woman had fallen from an original state of innocence. It is the only form of address Mr. BEAVER uses to introduce the CHILDREN to Mrs. BEAVER, and ASLAN uses these titles almost exclusively in addressing the children. TIRIAN, beholding FRANK and HELEN in ASLAN'S COUNTRY, feels awe in front of the Adam and Eve of his race.

SOPESPIAN — A scheming TELMARINE lord, friend of Lord GLOZELLE, counselor to MIRAZ, and his marshall of the lists. He is hacked to DEATH by King PETER (PC 13 and 14).

SOUNDS — Sounds (like SMELLS and TACTILE IMAGES) in the *Chronicles* often evoke entire experiences. The plucking of SUSAN's bowstring in PC brings back the old days more than anything that has yet happened, and "all the battles and hunts and feasts came rushing into [the CHILDREN's] heads together." The sound of the long breakers on RAMANDU's ISLAND is all-pervasive during the DAWN TREADERS' stay. The dark, flat voice of the WARDEN OF THE MARCHES OF UNDERLAND is a capsule version of the entire atmosphere of UNDERLAND. Sounds can also be treacherous and changeable. The QUEEN OF UNDERLAND tries to weave a spell with a monotonous thrumming on her lute that accompanies her SWEET, quiet voice; the HORNS of TASHBAAN sound exciting to SHASTA in the morning, but terrifying in the evening when he is shut outside the gates; and the "propputty-propputty" sound of HORSES' hooves on a hard road sounds like "Thubbudy-thub-budy" on dry sand. The sound of the trumpet announcing the Narnian war party is "clear, sharp, and valiant." Lewis contrasts it with the "huge and solemn" sound of the horns of Tashbaan, and the "gay and merry" sound of King LUNE's HUNTING horn.

SPARE OOM — Thought by TUMNUS to be the country from which LUCY came (the city being War Drobe). It is actually Professor Kirke's spare room, which houses the WARDROBE through which Lucy and her sister and brothers make their TRANSITIONS in and out of NARNIA (LWW 2).

SPEAR-HEAD, THE — NARNIA's guiding STAR, the brightest in its northern night sky. It is brighter than our North or Pole Star (LB 6).

[ASTRONOMY, NARNIAN.]

SPECTRES — Dreadful, terrifying GHOSTS, summoned by the WHITE WITCH to the slaying of ASLAN. As they leave the STONE TABLE, they pass by the place where LUCY and SUSAN are hiding. The girls experience the Spectres as a cold wind. Those Spectres remaining after the witch's defeat by Aslan's army haunt the battlefield for a while. (LWW 13, 15, and 17)

SPIVVINS — A weaker boy persecuted by the GANG at EXPERIMENT HOUSE, whom EUSTACE defends (in some unrevealed circumstance) by not saying anything about him, even when tortured (SC 1).

[AUTOBIOGRAPHICAL ALLUSION(S); SCHOOL(S).]

SPLENDOUR HYALINE — The royal galleon of NARNIA in its GOLDEN AGE, it is modeled after a swan. Luxuriously outfitted, it is nevertheless capable of being used in combat. Its silken sails and swanlike appearance may have suggested its name: *Hyaline* is from the Greek *hyalinos,* which means "glassy" or "smooth."[1] (PC 8 and HHB 5)

[*DAWN TREADER;* LITERARY ALLUSION(S).]

[1]Lewis may have had in mind a line from Milton's *Paradise Lost:* "On the cleer *Hyaline,* the Glassie Sea." (VII, 619)

SPRITES — Troublesome, terrifying, and hostile spiritual beings, present at the slaying of ASLAN (LWW 14).

Tied to a tree by the Calormenes and left to be killed by the followers of the false Aslan, Tirian is fed by three mice, two moles, and a rabbit. (LB 4)

SQUIRREL(S) — Flighty ANIMALS, they are the message-bearers of the *Chronicles.* These TALKING BEASTS bring news of RILIAN's safety to the owls, who are also messengers. Squirrels pour down from the TREES to help JILL, EUSTACE, PUDDLEGLUM, and Rilian out of the tunnel. PATTERTWIG is the only named squirrel in the *Chronicles.* (PC 15 and SC 15 and 16)

STABLE — A hut with a thatched roof, located on Stable Hill, the site of the deception of TASHLAN in LB. It is quite mysterious. TIRIAN walks around it and looks at it, noting that the stable DOOR seems to lead "from nowhere to nowhere"; but when he looks through the cracks in the wall, the assembly area can be seen. He concludes rightly that "the Stable seen from within and the Stable seen from without are two different places." DIGORY adds later that "its inside is bigger than its outside." Finally, LUCY compares it implicitly to the stable at Bethlehem, which also "had something inside it that was bigger than our whole world."

[MYTHOLOGY; PLATO.]

STAR(S) — The stars of NARNIA are actual beings, "glittering people with long hair like burning silver." They come into being at ASLAN's song of creation and shine and sing in "cold, tingling, silvery voices." RAMANDU was a star before the GOLDEN AGE OF NARNIA, and he is refreshed by the FIRE-BERRIES that grow on the SUN. His DAUGHTER, CASPIAN's wife and RILIAN's mother, has the blood of stars in her veins, which she passes on to the Narnian royal family. CORIAKIN the MAGICIAN was once a star, but he is being punished for some unknown failure by being put in charge of the Duffers. In an exchange between Ramandu and Eustace, we learn that what a star is made of and what a star is are two different things; the question of ingredients is far less important than the question of what kind of being a star is. The stars of Narnia look much larger than the stars of our world, because they are closer to their world than our stars are to our earth. The Centaurs of Narnia can read the stars and make predictions based on their positions. According to Roonwit, the stars never lie. At the end of Narnia, Aslan calls the stars home with the sound of the horn of the giant* Time, and they fall to earth until the skies are empty and black.

[Alambil; Aravir; Astronomy, Narnian; Creation of Narnia; Leopard, the; Reductionism; Tarva.]

STOCK RESPONSE(S) — The appropriate emotional and spiritual attitudes toward and feelings about true, good, and beautiful things, people, experiences, and events as well as toward and about false, evil, and ugly things, people, experiences, and events that every human being should have. Lewis felt very strongly, as did Plato, that children "must be trained to feel pleasure, liking, disgust, and hatred at those things which really are pleasant, likeable, disgusting, and hateful."[1] In the *Chronicles,* he wants to foster these stock responses in his readers. Digory's reiterated concern about Polly, about the sacredness of promises, about the wickedness of Mrs. Lefay, about cruelty to animals, about courage and cowardice and honor and justice show how well brought up he is in Lewis's sense: He has all the right stock responses. Conversely, Uncle Andrew becomes an evil MAGICIAN and a shriveled human being by trampling on and inverting his stock responses. He ignores his OATH to his godmother, he learns magic in unimaginable ways from awful people; he disrespects his body and loses his health—and all for the sake of so-called higher knowledge.

[1]*The Abolition of Man,* Chapter One, ¶10.

STONE KNIFE — This ancient stone knife is almost certainly the unseen knife being sharpened to kill EDMUND, and it is used by the WHITE WITCH to slay ASLAN. Later it appears on ASLAN'S TABLE, enshrined as a holy relic. One of the THREE SLEEPERS, ignorant of its sacredness, tries to use the knife in his quarrel with his comrades, but the moment he touches it they all fall into an enchanted SLEEP. The DAUGHTER OF RAMANDU then tells the voyagers that as long as the world lasts, the knife is to be kept in this place of HONOR, where each morning it catches the SUN's first rays. Like the cross of Christ, this ugly instrument of Aslan's DEATH has become a revered symbol of the atonement.

STONE TABLE — The great grim slab of gray stone, supported by four upright stones, upon which ASLAN is sacrificed. It is carved all over with strange lines and figures that might be the letters of an unknown language, and that cause a curious FEELING in onlookers. Its HISTORY is unknown, and it seems to have existed forever. According to the WHITE WITCH, it is the proper place for killing and the place where killing has always been done. Although its size is not mentioned it must be a low table, because the girls are able to kneel and still kiss Aslan's face as he lies on top of it. Unbeknownst to the White Witch, the Deeper MAGIC decreed before TIME began that the table would crack when a willing and innocent victim was killed "in a traitor's stead"—exactly the circumstances of Aslan's self-sacrifice for EDMUND's sake—and DEATH would begin to run backward. At the very moment the SUN comes up in NARNIA, the table breaks in two with a deafening noise. By the time of PC 7, the Stone Table is in the heart of ASLAN'S HOW (a mound erected over it) and the writing has all but worn away.

STORY, STORIES — The seven books of the *Chronicles of Narnia* are testament to the fact that Lewis loved stories and storytelling. The biggest difference between the New Narnians and the Old Narnians is the FAITH expressed in stories: MIRAZ thinks fairy tales are for CHILDREN and to be outgrown, while for CASPIAN the old stories are his salvation. Similarly, TIRIAN'S MEMORY of the old stories that told of mysterious children who came to NARNIA in times of crisis enables him to call on ASLAN and the SEVEN FRIENDS OF NARNIA for help.

Storytelling plays a large part in Narnian life, and in CALORMENE life as well. In fact, Calormene children are taught in SCHOOL to tell stories just as English children are taught to write essays—and the stories are of course much more interesting. ARAVIS tells her own story in grand Calormene style. After supper at CAIR PARAVEL, a blind poet comes forward to tell the story of the Horse and His Boy (SC 3). It is possible that stories may become distorted, as the story of the Lion Aslan has become, in Calormen, the story of a lion-shaped demon. When DIGORY hears

Aslan's "Well done," he knows that his victory over Jadis will be handed down in Narnia for a long time—indeed, Tirian asks Digory and POLLY if the stories about their journey to the GARDEN are true.

Some of the chronicles contain stories within stories within stories. The first half of PC is TRUMPKIN's telling of Narnian HISTORY since the reign of King Miraz, within which Doctor CORNELIUS relates the history of the TELMARINE conquest. In VDT, the spell for refreshment of the spirit is really a story that is so engrossing that the reader becomes part of it. On RAMANDU'S ISLAND, Caspian tells the DAUGHTER OF RAMANDU the story of Sleeping Beauty, which he must have been told by the Pevensie children. And HHB itself takes place within the context of the story of the last chapter of LWW.

Caspian is excited when he hears that EUSTACE comes from a round world—he has often heard stories of such worlds and longed to live in them but never believed they were real. Similarly, Tirian's vision of the children suddenly makes the old stories seem real. Finally, Digory discovers with a shock that fairy tales about EVIL* MAGICIANS are true when he encounters a modern magician, his Uncle ANDREW (who discounts the stories as "old wives' tales").

In the last paragraph of LB 16, Lewis turns the very *Chronicles* themselves into a metaphor for the meaning he has been trying to communicate through all his books. Even the finest STORY written or told by the finest storyteller is the barest prelude ("cover" and "title page" are the terms Lewis uses) to the Great Story being told through the lives of every person who has ever lived or will ever live. This Story will always be told by the Lion, now transformed into Jesus the Christ, whom Lewis believes is the Person behind these stories of every person in every world. SHASTA is right: Aslan is behind every story. And the Lion keeps his promise to LUCY: I will tell you the story for refreshment of spirit not only for years and years but for ever.

[LITERARY ALLUSION(S); MYTHOLOGY; VIOLENCE.]

STRAWBERRY — See FLEDGE.

SUICIDE — Two Narnian characters speak of suicide, both by a knife through the heart. First, in HHB 3, ARAVIS is desperate enough to avoid her forced betrothal to AHOSHTA and sad enough about her brother's DEATH in battle to want to join him in death that she attempts suicide but is prevented by HWIN, who reveals that she is a TALKING BEAST by saying, "O my mistress, do not by any means destroy yourself, for if you live you may yet have good fortune but all the dead are dead alike." When Aravis thinks she is so distressed that she is hearing things, Hwin interposes her head between the girl and her brother's dagger, and lovingly rebuking her "as a mother rebukes her daughter," gives Aravis reasons for living. In SC 9, PUDDLEGLUM is so distraught at his and JILL's and EUSTACE's accidentally eating Talking Stag (FEELING "as you would feel if you found you had eaten a baby") that he speaks of their having "brought the anger of ASLAN on" themselves and exclaims, "If *it was allowed,* it would be the best thing we could do, to take these knives and drive them into our own hearts" (emphasis added). Puddleglum's revulsion is a proper STOCK RESPONSE to an objective breach of RIGHT AND WRONG, but he errs in both understanding Aslan's reaction and also in even suggesting suicide. Lewis is clear that Christianity has no door marked "Exit"[1] for humans to choose and that even physician-assisted suicide is forbidden by God's law.[2] What is needed when one has such feelings of desperation is COURAGE, PRAYER, and the COMFORT of FRIENDS.

[1]*Surprised by Joy*, 171.

[2]*Letters to an American Lady*, 18 August 1958.

SUN — The celestial orb of NARNIA is a source of delight. In its mountains grow the FIRE-FLOWERS, and the birds of morning collect FIRE-BERRIES from its valleys. To the voyagers on the *DAWN TREADER*, the sun rising over the LAST SEA looks twice or three times as large as the sun they are used to. At the END OF NARNIA, the sun is a dying STAR and rises as a red giant, the moon rises too close to the sun, which absorbs it, and both are extinguished by the GIANT* TIME.

[ASTRONOMY, NARNIAN.]

SUSAN PEVENSIE — The second CHILD and eldest daughter of Mr. and Mrs. PEVENSIE; she is known as Queen Susan the Gentle in the GOLDEN AGE OF NARNIA, and also Queen Susan of the HORN. [MAJOR SPOILERS] As a young, beautiful, black-haired woman, she is so held hostage to her FEARS and to her desire to be thought mature and attractive that she is not included among those who are allowed to enter ASLAN'S COUNTRY.[1] Her fall from grace seems sudden and, to the extent that this appears so, shows an uncharacteristic lapse of style on Lewis's part. However, a careful rereading of her STORY shows that her fall is much better prepared for than some critics think. In the pre-1994 American editions Lewis changed the kind of ANIMAL she is hoping to see from rabbits to foxes, perhaps to suggest Susan's aspiration to foxHUNTING, an elitist sport. Susan's is one of the most important UNFINISHED TALES of the *Chronicles.*

Susan's first two lines in LWW reveal a girl aspiring to adulthood: She calls the professor "an old dear" and tells EUSTACE, "It's time you were in bed." When LUCY returns from her first visit to NARNIA, Susan uses more parental phrases: "What on earth," "Don't be silly," and "Why, you goose." When in Narnia herself, she is always practical and sensible; her first suggestion is that they don coats for the wintry weather from the WARDROBE, and she is worried about food. She does not find Narnia safe or fun, and though she too believes that TUMNUS ought to be helped, she'd rather not get involved; she wishes they had never come into this country.

One of her concerns is how safe it is to know ASLAN.[2] She receives a MAGICAL horn and a bow and arrows from FATHER CHRISTMAS as her gift from Aslan. Her tension eases on her walk through a reviving Narnia, but when she sees Aslan, she is so overcome with awe that she pushes PETER forward. As with Lucy, her FEELING that something awful is going to happen to Aslan keeps her from SLEEP, and the two agree to look for the Lion. Unlike Lucy, she worries that Aslan might be stealing away during the night and abandoning them to the power of the WHITE WITCH. Her care for others triumphs over her self-centeredness as she accompanies Aslan in his agony and watches his mistreatment at the hands of the witch and her forces. Becoming a little child again, she huddles with her sister. At the sound of the crack of the STONE TABLE, Susan is afraid to turn around until Lucy pulls her around. Like the risen Christ having to reassure his apostles that he is not a GHOST, Aslan breathes on her. Both sisters enjoy the romp with Aslan, but only Susan persists in asking what it all means. She is no longer enough of a simple child who can throw herself into whatever is happening; like an ADULT, she has to know. LWW ends with the picture of the mature young woman, much sought after as a bride but still disinclined to exertion in search of ADVENTURE.

In PC, Susan is the last of the four children to feel the pull into Narnia and it is quite unpleasant for her. She grows increasingly grouchy, going even so far as to consider refusing to cooperate further until she gets the kind of respect Lucy is getting. But when they finally get to the other side of the gorge, Susan admits to her sister that she believed Lucy that Aslan was commanding them to go another way. She confesses that "deep down inside" she knew her sister to be RIGHT. After the first phase of DANCING and feasting with BACCHUS, SILENUS, and the MAENADS subsides, Susan confides to Lucy that she would not have felt safe in the presence of all this wildness without Aslan at hand. As a girl moving into young womanhood, never an easy time, Susan is caught between the conflicting desires to be always a child and to be completely grown-up. She is neither here nor there, and the ecstatic side of life would be too much for her to deal with, were it not for Aslan, in whose presence all REVELRY has its place.

[1]This is not to say, as some critics have maintained, that she is lost forever. Lewis intends only to explain how it is possible to

reject the JOY that comes from being in Narnia and also to illustrate one way of doing so. It is a mistake to think that Susan was killed in the RAILWAY ACCIDENT at the end of LB and that she has forever fallen from grace. It is to be assumed, rather, that as a woman of twenty-one who has just lost her entire family in a terrible crash, she will have much to work through; in the process, she might change to become truly the gentle person she has the potential for being (see *Letters to Children,* 51 and 69). Lewis gives the impression of harsh and final judgment of Susan by the SEVEN FRIENDS OF NARNIA, an ungenerosity and superficiality he condemns in his other writings.

[2]This is a reflection of one of Lewis's main concerns. "I am a safety-first creature. Of all arguments against love none makes so strong an appeal to my nature as 'Careful! This might lead you to suffering,'" said Lewis in *The Four Loves,* 168. Lewis wanted to avoid PAIN and interference at all costs. Perhaps one of the reasons he is able to write so well of Susan's character is that he experienced in his own personality her reluctances, fears, and desire to be thought adult.

SWANWHITE — Queen of NARNIA sometime before[1] the Hundred Years of Winter. Her beauty was so radiant that any pool she used as a mirror would reflect her face for a year and a day. (LB 8)

[MAGIC; UNFINISHED TALE(S).]

[1]This is contradicted by Lewis's outline of Narnian HISTORY (*The Land of Narnia,* 31), in which he states that Swanwhite was ruling in 1502 N.Y.

SWEET(NESS) — A pleasurable quality of SOUND or SMELL or taste that is often a clue to whether an experience is RIGHT OR WRONG. The taste of TURKISH DELIGHT is too sweet, as are the trilling voice of the QUEEN OF UNDERLAND and the incense she burns in her fireplace. So too is the ring of the GOLDEN BELL in MN. One of the few good sweetnesses in NARNIA belongs to the waters of the LAST SEA.

SWEET WATERS — See ASLAN'S BREATH; LAST SEA

T

TACTILE IMAGE(S) — Lewis's use of touch, SMELL, and SOUND in the *Chronicles* helps to make the CHILDREN'S fabulous ADVENTURES much more real. LUCY'S TRANSITION between the WARDROBE and NARNIA is marked by the feel on her skin of prickly fir branches turning into soft fur coats. SC is filled with such images: JILL'S Narnian clothes are not only nice looking but nice feeling; GLIMFEATHER'S feathers "felt beautifully warm and soft"; the children's beds in the MARSH-WIGGLE'S wigwam are soft and warm in contrast to the cool, damp night air; and at HARFANG Jill's enjoyment of the GIANT towels after her bath is especially sensual. Royal ROBES look and feel splendid. DIGORY'S bath in the mountain stream in MN is compared with a sea-bath, an AUTOBIOGRAPHICAL ALLUSION.

TALKING BEAST(S) — The ANIMALS of NARNIA who have been given the GIFT OF SPEECH and dominion over the DUMB BEASTS by ASLAN. They are not the same size as ordinary animals, the smaller ones being larger (talking MICE, for example, are two feet tall) and the larger ones being smaller. Their faces shine with intelligence. Although we may assume that Talking Beasts grow old and die (Mr. and Mrs. BEAVER are long dead by the time of VDT), they are not shown to age.

Talking Beasts are created by Aslan from the ordinary animals. He goes among them and touches pairs of each type of animal by the sides of their noses; these animals instantly follow him. In his commands to Narnia to awake, he commands them to "Be Talking Beasts." The animals know ASLAN'S NAME intuitively and pledge their OBEDIENCE. In return, Aslan gives them their world, their selves, and himself.

Although some are later corrupted (notably, SHIFT), the Talking Beasts are originally innocent. Lewis implies this innocence by having them confuse even the sound of the words "an EVIL" into "a Neevil." Their natural CURIOSITY about Uncle ANDREW and the desire to exercise their newly made bodies move them to pursue the fleeing elderly gentleman in a chase that is a deliberate reversal of an English foxhunt, with all the appropriate shouts. Some (with a degree of truth) think he is the "Neevil" Aslan is looking for.

Being a Talking Beast is both an HONOR and a responsibility. When they gnaw through the ropes that bind Aslan to the STONE TABLE, the mice are honored for their help by being transformed into Talking Beasts. REEPICHEEP takes himself quite seriously, and when CASPIAN slurs his mousehood he responds by reminding the KING of his coronation OATH to be a good lord to the Talking Beasts of Narnia. Aslan warns the newly created Talking Beasts that they must abstain from their former ways as DUMB BEASTS or lose the gift of speech. When this does in fact happen to GINGER the CAT, the other animals are terror-stricken.

However, it is possible to be redeemed: The LAPSED BEAR OF STORMNESS, who has gone wild, is reformed by CORIN THUNDER-FIST.

The Talking Beasts are in many ways similar to humans. Causing PAIN to animals is abhorrent to Narnians (and to Lewis), and cruelty to Talking Animals is the worst of all. The evil side of the CALORMENE character is revealed by the way in which animals are treated in LB. Although it is permitted to HUNT dumb beasts for food (there are no vegetarians in Narnia), to eat a Talking Beast is akin to cannibalism. EUSTACE, JILL, and PUDDLEGLUM are horrified to discover, halfway through a meal, that they have been eating Talking Stag.[1]

Lewis uses the Talking Beasts as hieroglyphs, word-pictures that embody various human attributes (Reepicheep, for example, is a hieroglyph of COURAGE). By using animals, Lewis can make graphic statements about the human condition that would be much more complicated to express in human terms. By portraying Talking Beasts that live in aid and FRIENDSHIP to human beings, Lewis reminds us that we are indeed part of the natural world, and not separate from it as modern science and TECHNOLOGY might have us believe.

[CREATION OF NARNIA; EVIL; HIERARCHY; VIOLENCE.]

[1]Each reacts to this knowledge according to how long he or she has been in Narnia. Jill, a newcomer, is sorry for the animal and outraged at the cruelty of the GIANTS; Eustace, with one Narnian ADVENTURE under his belt and Reepicheep for his friend, says it is like murder; Puddleglum, a native Narnian, feels it as deeply as we would feel if we had eaten a baby. See SUICIDE.

TARKAAN — A CALORMENE great lord; the feminine form is "Tarkheena." Tarkaans are accustomed to commanding groveling OBEDIENCE from their social inferiors in the Calormene HIERARCHY. *Tarkhan* (alternate spelling: *Tarkaan*) was a rank of minor nobility among medieval Mongols.[1]

[1]The reference is supplied by John Singleton.

TARVA — One of two noble planets (the other is ALAMBIL) that every two hundred years pass within one degree of each other in the Narnian night sky. Doctor CORNELIUS surnames Tarva "The Lord of Victory" (PC 4 and 6).

[ASTRONOMY, NARNIAN.]

TASH — The chief GOD of CALORMEN, invoked by worshipers as "Tash the inexorable, the irresistible." He is a cruel, bloodthirsty GOD with a taste for human sacrifice. Vaguely man-shaped, Tash has a vulture's head, four arms, and twenty fingers that end in cruel, curving talons. A deathly smoke surrounds him and grass wilts under his step. His voice is the "clucking and screaming" of a "monstrous bird." Lewis derived the god's name from *tash* or *tache,* a Scottish word that means "blemish, stain, fault, or vice." Many Calormenes (with the notable exception of EMETH) pay only lip service to their god, and when he actually shows up at the STABLE they are terrified. He almost eats the atheist GINGER, does eat the despicable SHIFT, and is about to do something equally terrible to RISHDA when he is banished to his "own country" (presumably HELL) by one of the seven KINGS and queens (presumably the High King PETER). While this seems to indicate that Lewis had Tash in mind as a picture of the devil, in HHB he is merely the opposite of ASLAN's qualities rather than of Aslan himself. The appearance of Tash in NARNIA is perhaps the single most terrifying scene in the *Chronicles* (LB 12).

[DEPRAVITY; TASHLAN; UNIVERSALISM.]

TASHBAAN — The capital city of CALORMEN and the seat of the TISROC's rule; its name is derived from the Calormene GOD* TASH, and *baan*.[1] Tashbaan is located on an island in the middle of a broad river, and circles of buildings rise from the mean commercial district at the bottom, to the upper-class homes, to the Tisroc's palace at the very top. The only traffic rule is that "everyone who is less important has to get out of the way for everyone who is more important." Interestingly, from her vantage point in the GARDEN in LB, LUCY can see Tashbaan down below in the real NARNIA: It is a great city. Thus Lewis must feel that there is something good in the Calormene culture (as symbolized by its chief city) that allows it to pass into eternity (LB 16).

[1]Arabic for "house of."

TASHLAN — The word is a combination of the names "TASH" and "ASLAN," a lie invented by SHIFT and the CALORMENES (it could be the invention of RISHDA) in LB (3, 9, 11, and 15). Lewis allows that some Narnians may have an "honest FEAR" of Tashlan that is not prompted by treachery, but EMETH, the noble Calormene officer, is outraged by the equation of Tash, whom he loves, with Aslan, whom he hates.

[DEPRAVITY; END OF NARNIA; ESCHATOLOGY; UNIVERSALISM.]

TECHNOLOGY — This is a neutral term, referring to the PRACTICAL uses of science. The negative term, "technologism," refers to the increasing separation of humanity from nature brought on, in Lewis's view, by the application of science to all aspects of human life; this attitude is the opposite to ECOLOGY. In LWW, EDMUND, enchanted by EVIL* MAGIC, distracts himself from his real situation with power fantasies of how much better NARNIA will be when he is KING and he can make "decent" roads, build his palace with a cinema, own a fleet of automobiles, and install a railway system. EUSTACE, in VDT, is a perfect product of EXPERIMENT HOUSE; he sees all that is best in Narnia—that is, all that is natural—as primitive and second-rate. He calls the *DAWN TREADER* a "blasted boat" and says that in "civilized" countries "ships are so big that when you're inside you wouldn't know you were at sea at all." On RAMANDU'S ISLAND, Eustace remarks that on his world "a STAR is a huge ball of flaming gas"—a great example of REDUCTIONISM. RAMANDU corrects this assumption with the reply, "Even in your world . . . that is not what a star is but only what it is made of."

It is this preoccupation with what things are made of—the sum of the parts—as opposed to the appreciation of the whole that Lewis finds particularly abhorrent. The evil people of the *Chronicles* are characterized by their exaggerated concern for the practical, in the sense of the usefulness of things and people *to them*. At the sight of the growing LAMP-POST, Uncle ANDREW excitedly exclaims that "the commercial possibilities of this country are unbounded" (MN 9). In LB, the triumph of technology over

It is Aslan's final coming. He bounds down the rainbow cliffs, a herald of power and glory. (LB 16)

nature leads to the END OF NARNIA. SHIFT intends to sell the TALKING BEASTS and other ANIMALS into SLAVERY, and to use that income to build "roads and big cities and schools and offices and whips and muzzles and saddles and cages and kennels and prisons" (LB 3).

TELMAR, TELMARINE(S) — Telmar is a land far beyond the WESTERN WILD (the great forest lies between Telmar and LANTERN WASTE), from whence the Telmarines came into NARNIA. The Telmarines, or New Narnians, are the offspring of marauding human pirates who blundered into Telmar in 460 N.Y.[1] through a cave—one of the last MAGIC connections between our world and that world. There they married native women, and in 1998 N.Y.[2] their fierce, proud descendants invaded a disordered Narnia and conquered it. The first Telmarine KING of Narnia was CASPIAN I, and under his rule the Telmarines went about "[SILENCING] the beasts and the TREES and the fountains" and driving away the DWARFS and FAUNS. Throughout their rule Old Narnians must hide their identities, and the old STORIES that tell the truth about ASLAN and Old Narnia are largely forgotten or ridiculed as "old wives' tales." The Telmarines' hatred of nature is complemented by their superstitious FEARS, and they call the GREAT WOODS the Black Woods because they imagine the woods to be filled with GHOSTS. [PC SPOILERS] Captured by CASPIAN X's army, the Telmarines make a fuss about wading across a river because they are afraid of running water. At last all are compelled either to accept Aslan's restoration of Narnia or to go to a home back on earth that has been prepared by Aslan. Younger Telmarines want to stay on in

Narnia, while the older ones are sulky and suspicious. That some Telmarines do stay is evidenced by the fact that Lord RHOOP, in VDT 12, identifies himself as a "Telmarine of Narnia." Some Telmarines eventually do follow the Pevensie CHILDREN back through the DOOR in the air, and they are deposited on a Pacific island.

[ECOLOGY; IMAGINATION; TECHNOLOGY; TRANSITION(S).]

[1]*The Land of Narnia,* 31. Hence the meaning of their name: *tel* means "earth" and *mar/ines* means "sea/sailor." (Suggested by Dr. Joe R. Christopher in "An Inklings Bibliography," *Mythlore* 27.) But see POLLY PLUMMER, n. 1.

[2]*The Land of Narnia,* 31.

TERM-TIME — The time when SCHOOL is in session, especially the beginning of the school term, when CHILDREN must return from their holidays or their vacations. In the *Chronicles,* term-time[1] is a metaphor for the unenjoyable seasons of life; thus the Pevensie children are fortunate to be whisked away to NARNIA while on their way back to boarding school, only to be returned, of course, at the end of their ADVENTURE to face the summer term ahead of them. It is JOY indescribable to hear from ASLAN at the end of LB (16), "The term is over: the holidays have begun."[2]

[1]British schools have three terms, each about twelve weeks in length. Autumn term runs from about September 20 to December 18; spring term runs from about January 14 to early April; and summer term runs from the end of April to mid-July. Oxford and Cambridge terms are eight weeks in length. Oxford terms are named Michaelmas (from the Feast of St. Michael, September 29), Hilary (from the Feast of St. Hilary, January 13), and Trinity (from Trinity Sunday, the first Sunday after Pentecost). Cambridge terms are named Michaelmas, Lent, and Easter. It would not, then, be surprising for the Pevensie chil-

dren to be dressed in winter clothing at the beginning of the summer term (see the very first and very last paragraphs of PC) because summer term would normally begin in late April, usually a cold month in ENGLAND.

[2]See SLEEP for a discussion of parallels between term-time and sleep, and vacation-time and waking.

THEODOLINDS — See HOUSE OF THE MAGICIAN.

THREE BULGY BEARS — TRUFFLEHUNTER's nearest neighbors, who live to the east of him in the northernmost part of ARCHENLAND. Though good-hearted, these TALKING BEASTS give first place to food and habits of COMFORT. When they attend the great council, the bears want to eat first and counsel later. However, they are not totally without ambition. When King PETER is to face MIRAZ in single combat, they insist on their long-standing right to have one of their number be a marshall of the LISTS. King Peter appoints the oldest bear to this office even though, true to his nature, the bear is prone to sucking his paw and falling asleep. Later on, CASPIAN confirms the bear permanently in his hereditary office of marshall of the lists. (PC 6, 7, 13, 14, and 15)

THREE SLEEPERS — ARGOZ, REVILIAN, and MAVRAMORN, three of the SEVEN NOBLE LORDS found in and awakened from an enchanted SLEEP on RAMANDU'S ISLAND. (VDT 13, 14, and 15)

TIME[1] — Time in NARNIA and time in ENGLAND behave quite differently: While time in England flows on in the usual manner, any number of years may have passed in

Narnia in the same amount of "time." From its creation to the LAST BATTLE, Narnia has a total HISTORY of 2555 Narnian years (N.Y.). The same amount of time in England spans only forty-nine earth-years, from 1900 to 1949. Furthermore, there is no direct correspondence: One earth year does not necessarily equal a corresponding number of Narnian years.

Even in our world time passes much more slowly for CHILDREN and YOUTH than it does for ADULTS, and this attitude is evident in the Narnian world of LWW, in which it is always winter but Christmas never comes. To young children, each winter must sometimes seem like a hundred years.

Discrepancies between earth time and Narnian time lead to other interesting speculations, especially PETER'S observation in PC that the children's return to Narnia after so long (even though it was only a year in England, it was 1303 years in Narnia) must be something like what it would be like for the Crusaders or ancient Britons to find themselves alive and well in the modern world.

There are two timeless places in Narnia: the WOOD BETWEEN THE WORLDS, which is almost outside of time; and ASLAN'S COUNTRY, which is eternal. The END OF NARNIA also signals the end of time; ASLAN comments that FATHER TIME'S name was "time" when he was SLEEPING (the same period that Narnia was awake and living), but now that Narnia has ended he will have a new name.

[1]See *The Land of Narnia*, 31.

TIRIAN — The last KING of NARNIA, the seventh in descent from King RILIAN. In his early twenties, he is well developed, with a scanty beard, blue eyes, and a "fearless, honest face." Although he has the nobility and HONOR of all of the kings of Narnia, he cannot escape his country's overwhelming atmosphere of decline. His emotions are not under control, and he swings from elation to depression to remorse to anger. Lewis would like us to see Tirian as the noble king in the face of the inevitable Twilight of the Gods, but this is not made as clear as it might be.

Tirian dies at the LAST BATTLE. Once inside the STABLE* DOOR, he is surprised to discover the SEVEN FRIENDS OF NARNIA. The climax of Tirian's entry into ASLAN'S COUNTRY is his reunion with his dead father, ERLIAN, which is perhaps the most genuinely emotional scene between an ADULT and a YOUTH[1] in the *Chronicles.*

[1]See AUTOBIOGRAPHICAL ALLUSION(S).

TISROC — The hereditary title of the KINGS of CALORMEN; it is never mentioned without the obligatory "may-he-live-for-ever." Lewis probably derived "Tisroc" from "Nisroch," the Egyptian god in Edith Nesbit's *The Story of the Amulet.*[1]

[1]E. Nesbit, *The Story of the Amulet,* in collection with *Five Children and It* and *The Phoenix and the Amulet* (London: Octopus Books, 1979), 495.

TOFFEE TREE — A TREE about the size of an apple tree, very dark-wooded, with whitish, papery leaves and datelike fruit. It grows overnight from one of POLLY'S nine toffees (all she and DIGORY have to eat on their journey to the GARDEN in the west), which Digory plants as an experiment. (MN 12 and 13)

TRANSITION(S) — Human CHILDREN are called into NARNIA in times of crisis, and only at ASLAN's bidding or with his tacit permission. It is, however, possible to get in accidentally (as with the TELMARINES) or by being attached to someone else (as with Jadis and Uncle ANDREW). The transition from our world to Narnia and back again is accomplished in a number of ways. In PC 15, Lewis explains that the Telmarines first made their way from earth to Telmar through a cave, which is one of the last "magical places . . . one of the chinks or chasms between worlds." In LWW, the Pevensies go back and forth between ENGLAND and Narnia through the WARDROBE. Even though in Narnia they are KINGS and queens and in England they are still children, the transition is not abrupt. LUCY, for example, feels[1] the branches of the *fir* TREES turning into *fur* coats, a transition that is sensual for her and visual for the reader. In PC 1, PETER, SUSAN, EDMUND, and Lucy are literally pulled by MAGIC from a train station in England right into Narnia. On the return trip (PC 15), Aslan makes a DOOR in the air, which they and the Telmarines walk through. In VDT 1, Lucy, Edmund, and EUSTACE enter Narnia through a painting of the *DAWN TREADER:* First the picture looks real, then the wind blows, next noises are heard, then water splashes, and there they are on the real ship in Narnia. Their return to England is accomplished through another door that Aslan makes in the sky (VDT 16). In SC 1, Aslan calls JILL and Eustace into Narnia through a doorway behind EXPERIMENT HOUSE. As soon as they step through it, England disappears. On the return, Aslan leads the children and CASPIAN through the woods, where the wall of Experiment House suddenly appears before them. Aslan makes a gap in the wall through which they can see everything they had left behind, and in this transitional place the GANG can see them as well (SC 16). In MN, most of the transitions are accomplished magically, via the magic RINGS. But in the final return to England, there is no need to use them—it is the swirling vision of Aslan himself that sends POLLY and DIGORY back (15). In LB 4, there is a very different transition. TIRIAN and the SEVEN FRIENDS OF NARNIA actually appear to one another "as if in a DREAM," and Eustace and Jill simply appear before Tirian, without explanation. The final transition is not from England to Narnia but from Narnia to the real Narnia, ASLAN'S COUNTRY, into which all the good creatures of Narnia enter through the STABLE door. Once on the other side of the door, they can see the END OF NARNIA. And in the GARDEN of the west they can see that all that is real in Narnia and England is connected, and it is only necessary to walk in order to get from one place to another.

[1]See TACTILE IMAGE(S).

TREE OF PROTECTION — A TREE grown by DIGORY from the seed of the silver apple of the GARDEN of the west to protect NARNIA from the WHITE WITCH. When Digory buries the core from one of its apples outside the Ketterley house in London, a tree grows up overnight. Deep down inside, the tree is one with its Narnian parent, and it moves its limbs when Narnian winds blow. The tree is ultimately destroyed by a gale, and Digory has its wood made into a WARDROBE. (MN 14 and 15)

TREE-PEOPLE — Beings that are part tree, part human.[1] As TREES they look only vaguely human, but when some good MAGIC calls them to life, they look like leafy, branchy GIANTS and giantesses.[2]

[ASLAN'S VOICE; CREATION OF NARNIA; DRYAD(S); HAMADRYADS; SILVANS; WOOD PEOPLE.]

[1]They are modeled on the woods at Helm's Deep in J. R. R. Tolkien's *The Lord of the Ring: The Two Towers* (New York: Houghton Mifflin, 1965), 146–147, 151. On the other hand, Tolkien's Treebeard is modeled on his friend C. S. Lewis. (See Humphrey Carpenter, *Tolkien* [New York: Houghton Mifflin, 1977] 194.)

[2]For a more detailed discussion of male and female trees, see DRYAD(S).

TREE(S) — C. S. Lewis shares with J. R. R. Tolkien a deep love of trees. The trees of NARNIA are truly alive; at ASLAN'S command they become waking trees. Trees of Narnian countries include many species of English trees, such as oaks, hollies, beeches, silver birches, rowans, firs, pines, sweet chestnuts, and apple trees. Most seem to be good—the Oak is present at the first council—but some are capable of wrongdoing.

[DRYAD(S); ECOLOGY; HAMADRYAD(S); SILVANS; TECHNOLOGY; TREE-PEOPLE; WOOD PEOPLE.]

TRINITY — The central revealed mystery[1] of the Christian PROFESSION OF* FAITH: Father, Son, and Spirit. In the entire *Chronicles* there is only one specific reference to the Trinity. It comes in HHB 11 when SHASTA, weary and alone, pours out his STORY to the Large Voice walking beside him in the foggy night. In answer to Shasta's question, "Who *are* you?" the Voice gives a threefold "Myself" in three different tones of voice: (1) "very deep and low so that the earth shook"; (2) "loud and clear and gay"; and (3) an almost inaudible but also all-encompassing whisper. The first suggests the power and the GREATNESS OF GOD, who in NARNIA is represented by the EMPEROR-BEYOND-THE-SEA. The second reveals the Eternal Word and highlights his eternal YOUTH and JOY. The third emphasizes the subtle yet all-pervasive activity of the Holy Spirit, ASLAN'S BREATH.

This revelation removes Shasta's FEAR of being devoured or of being in the presence of a GHOST; the "new and different trembling" he now feels is the fear of God, properly so called, the terror and delight that are the hallmarks of a genuine experience of the NUMINOUS as Lewis sees it.

[1]It is called a mystery, not with the connotation of being *kept* secret, but because human minds have very few handles by which to grasp it and because they are too small to comprehend it. The legend of St. Augustine and the Child explains this inadequacy thus: After a particularly successful morning of theologizing, this late-fourth to early-fifth-century bishop of the coastal North African city of Hippo decided to rest his mind by walking along the seashore. There he saw a CHILD working quite hard at an excavation in the sand. The boy then began to run between the sea and the hole, carrying tiny buckets of water to the hole. The theologian asked the child what he was doing. "I am trying to pour the sea into this hole," he answered. The bishop laughed aloud and told the boy how impossible this task was. The child responded, "So too is your boast that you have understood the meaning of the Trinity," and he disappeared, leaving a much humbled thinker.

TRUFFLEHUNTER — A talking badger with a large, friendly, intelligent face. He is another of Lewis's solid, sensible, and, above all, loyal characters (Lewis loved Mr. Badger of Kenneth Grahame's *The Wind in the Willows*). Trufflehunter knows his place and defers to Doctor CORNELIUS as a "learned man." His role in life is also clear to him: Badgers, he tells GLENSTORM, have the duty to remember (see MEMORY), just as CENTAURS are meant to watch the SIGNS in heaven and earth. When PETER calls Trufflehunter the best of badgers, he explains that being a beast, he does not change; and being a badger, he holds on.

True to his nature, Trufflehunter tries to "badger" his companions into faithfulness and RIGHT conduct. Trufflehunter's PROFESSION OF FAITH prompts him to urge everyone to be patient in the certainty that ASLAN will come to their rescue. An argument between the badger and the DWARF* NIKABRIK escalates until the latter attacks Trufflehunter. Nikabrik killed, the badger is embraced and kissed by Peter. From then on he sits quite close to Peter and never takes his eyes from him—another indication of Trufflehunter's deep sense of loyalty.

Trufflehunter is HONORED with the Most Noble ORDER of the Lion by CASPIAN, who will later (in VDT 16) ask him to select a KING in his place. The white patches on Trufflehunter's cheeks are the last things LUCY sees in NARNIA. He is among the blessed in the GREAT REUNION.

[BIBLICAL ALLUSION(S); DANCE.]

TRUMPKIN — A Red DWARF; he is the Pevensie CHILDREN's DLF throughout PC, CASPIAN's regent in VDT 16, and the old, deaf lord chancellor of SC 3 and 16. Trumpkin is a medium-sized Dwarf, about three feet tall. Stocky and deep-chested, he has an immense beard and red whiskers that frame a beaklike nose and twinkling eyes. His HUMORous way of speaking is full of alliterative expletives, which include "horns and halibuts," "whistles and whirligigs," "giants and junipers," "bottles and battledores," "bilge and beanstalks," "cobbles and kettledrums," "wraiths and wreckage," "weights and water bottles," and "crows and crockery." Although as an Old Narnian he might be expected to have FAITH in ASLAN, for much of PC he is an agnostic. He is, however, brave and loyal. Trumpkin is above all a realist; he doesn't believe anything unless he has seen it with his own eyes.[1] Even when Trumpkin finishes telling his STORY to the four CHILDREN, he does not make the connection between them and help from Aslan. He is persuaded to believe in them only when he is bested at archery by SUSAN; and (like BREE in HHB) he believes in Aslan only when the two are face-to-face and the Lion (who likes the Dwarf very much) grabs him in his mouth, tosses him into the air, catches him, and stands him up again. After the defeat of MIRAZ, Caspian raises the Dwarf to the Most Noble ORDER of the Lion. Trumpkin passes the JUDGMENT into ASLAN'S COUNTRY and is present at the GREAT REUNION.

[1]This seems to be a basic trait of Dwarfs. It is taken to its extreme by the Dwarfs at the STABLE, who see only what they want to see and can no longer see what is really there.

TUMNUS[1] — A flute-playing FAUN who is approaching middle age and becoming stout, he has the characteristics of his MYTHOLOGICAL race: the legs of a goat, the upper body of a man, curly hair, and two little horns on his head. He is about the same height as SHASTA, probably about four feet tall (the same height as a good-sized DWARF). He is the first Narnian that LUCY—and the reader—meets in

LWW, and she politely calls him Mr. Tumnus. For not turning her over to the WHITE WITCH he is eventually arrested and turned into a stone statue in the courtyard of the witch's castle. Revived by ASLAN, he DANCES for JOY with Lucy. He is the first-named FRIEND to be rewarded and HONORed at the coronation of the KINGS and queens. In HHB 4, he is a loyal Narnian who hates every stone of TASHBAAN. It is his idea to stock the *SPLENDOUR HYALINE,* to fool RABADASH, and to escape. [LB SPOILER] He enters into ASLAN'S COUNTRY at the judgment and discloses to Lucy the mystery of the glory of the within-ness of the GARDEN and NARNIA: "The further up and further in you go, the bigger everything gets" (LB 16).

[1]The origin of his name is not clear, although it is a Latin word of some sort. It may be from *tumulus,* meaning "hill," as Tumnus lives in hilly country. Nancy-Lou Patterson suggests that *Tumnus* comes from *Vertumnus,* the Latin word for "a shape-shifting sylvan divinity of Roman religion" (*Mythlore* 27).

TURKISH DELIGHT — EDMUND's favorite confection, it is a SWEET candy made with sugar, fruit juices (such as orange, lemon, lime, and strawberry), rosewater, gelatin, and slivered nuts (such as almonds and pistachios) and dusted with powdered sugar (you can find a recipe on the Internet); it is often served as a special Christmas treat. Edmund GREEDily eats several pounds of the WHITE WITCH's enchanted Turkish Delight in LWW, and as a result he falls under her power. (4, 7, 9, and 11)

TWO BROTHERS OF BEAVERSDAM — TELMARINE lords, rulers at BEAVERSDAM under CASPIAN IX. MIRAZ imprisons them under the false charge of madness. (PC 5)

U

UNCLE ANDREW — See Andrew Ketterley.

UNDERLAND — [SC SPOILERS] The generic term for all the territory that lies beneath Narnia (SC 10–14). It is bordered at the top by the Marches; the Deep Realm is very much farther down; and Bism,[1] the bottommost realm, is one thousand fathoms beneath them. The inhabitants of Underland, the Earthmen, call the land above them Overworld or Upperworld and refer to the people there as Updwellers. Underland is a dark, cavernous world, sad and quiet and held in enchantment by the queen. Mosses and treelike shapes grow in the tunnels, and dragon- and batlike creatures sleep, waiting to wake at the end of Narnia. The capital city of Underland is a great seaport situated on the Pale Beaches, and all the outlets into Overworld are located on the other side of the Sunless Sea (except the escape tunnel that is dug directly under the site of the Great Snow Dance). The Earthmen, who originally came from Bism, call Underland the Shallow Lands (which, in their view, it is).

Bism, in contrast to Underland, is a bright land of molten rocks. A river of fire inhabited by salamanders flows through Bism, and its banks are covered with fields and groves of a tropical brilliance. Edible gems of all sorts grow like fruits in Bism—rubies grow in bunches like grapes, diamonds like oranges. After the liberation of Underland, the Narnians keep the entrance open. It becomes quite an attraction, and on hot summer days Narnians descend to the Sunless Sea with ships and lanterns to sail and sing and enjoy themselves.

[PLATO.]

[1]Appropriately, *bism* is an obsolete form of *abysm*, Greek for "bottomless pit."

UNFINISHED TALE(S) — In LB 8, JEWEL tells JILL of SWANWHITE the Queen, MOONWOOD the Hare, King GALE, and "whole centuries in which all NARNIA was so happy that notable DANCES and feasts, or at most tournaments, were the only things that could be remembered, and every day and week had been better than the last." These are just a few of the untold STORIES of Narnia that Lewis encouraged CHILDREN who wrote him about more Narnian chronicles to write themselves: Of Denise on 8 September 1962 Lewis asks, "Why not write stories for yourself to fill up the gaps in Narnian history? I've left you plenty of hints—especially where LUCY [*sic*] and the Unicorn are talking in *The Last Battle.* I feel I have done all I can!"[1] The last two paragraphs of HHB 15 leave unfinished the stories about how CORIN became Prince Corin Thunder-Fist and why RAM the Great was the most famous of the KINGS of ARCHENLAND. There is also the question of what happened to SUSAN after the DEATH of her entire family in the RAILWAY ACCIDENT.

[1]See *Letters to Children* for similar letters to Jonathan on 29 March 1961, and to Sydney on 14 February 1962.

UNICORN — A MYTHOLOGICAL beast with a single horn in the center of its head. It variously symbolizes purity, chastity, and even the word of God as brought by Jesus Christ. According to legend a Unicorn has the legs of a buck, the tail of a lion, and the body and head of a HORSE, although in art it is most often depicted as a white horse (as it is in the *Chronicles*). Also according to legend, it has a white body, a red head, and blue eyes; the horn is white at the bottom, black in the middle, and red at the end. It is fierce in battle. Unicorns are mentioned several times in LWW (12, 13, and 16), and the noble and delicate JEWEL is present throughout LB. In contrast to the classical description, Lewis's Unicorns have indigo blue horns.[1] One is present in the group that forms a half-circle around ASLAN before PETER's first battle, and at least two are sent to rescue EDMUND. Like horses, Unicorns carry smaller creatures into battle. They use their horns as weapons.

[1]Lewis has his own wonderful "vision" of Unicorns. In *The Great Divorce,* Chapter VIII, last ¶, he describes a herd of Unicorns: "twenty-seven hands high the smallest of them and white as swans but for the red gleam in eyes and nostrils and the flashing indigo of their horns."

UNIVERSALISM — The belief that all religions are basically the same and thus one is as good as another. Two scenes in the *Chronicles* raise in the mind of some older readers the question of whether or not C. S. Lewis was a universalist. Did he believe that all roads lead ultimately in the same direction: eternal happiness for everyone? Or did he favor Christianity over the other world religions and believe that it is possible for a person to choose to be eternally unhappy and therefore be in some kind of HELL?

In the first of these scenes (HHB 15), after ASLAN turns RABADASH into an ass, the Lion declares that because the CALORMENE prince has invoked his GOD* TASH, any release that the prince will experience from this enchantment will take place in the temple of Tash. (This does not mean that Lewis is teaching that Tash is bestowing this gift; all who know the story will know that Aslan is giving this gift because Aslan is real and Tash is not.) Aslan warns Rabadash not to stray more than ten miles from this temple or he will return to the form of an ass forever—there will be no second chance. Though universalism can be viewed as a belief in any number of chances, this passage does not disabuse some readers of the notion that, since Lewis permits even one more chance, he is relativizing the traditional view that life before DEATH is an unalterable experience—that "as a tree falls, so shall it lie."

As further and much clearer evidence of Lewis's universalism, some readers point to the dialogue between Aslan and the Calormene soldier EMETH in LB 15. Aslan accepts Emeth's devotion to Tash as devotion really paid to himself and addresses Emeth as "son," "child" (twice), and "beloved." Even though Emeth has not known Aslan except as the lion-shaped demon that Calormenes believe to be the power behind NARNIA, Aslan has known and loved Emeth. When Emeth tries to deflect the Lion's welcome by reminding Aslan that he has been Tash's servant, Aslan tells him that he has drawn that service to himself. Emeth asks if SHIFT's lie about the identity of Tash and Aslan is really true. Lewis carefully uses the adverb "wrathfully" to describe Aslan's response and details the radical differences between Aslan and Calormen's chief god. Aslan asks Emeth if he understands this explanation and Emeth's ambiguous reply suggests both that he does not comprehend (that is, *fully* understand) and that he guesses the *meaning* of what the Lion has said even if he doesn't grasp all the *facts*.[1] Emeth goes on to ask what has been the meaning of his lifelong pursuit of Tash. Aslan responds, "Beloved, unless thy desire had been for me thou wouldst not have sought so long and so truly. For all find what they truly seek."

SHASTA, the hero of HHB, is in some way a first sketch of Emeth: Under the "influence" of Tash, Shasta experiences Aslan's help before he knows ASLAN'S NAME. But Aslan is able to do little with the unrepentant ANDREW Ketterley in MN 14 except to render the man senile. Aslan is able to do nothing for the renegade DWARFS in LB 13.

So, far from being the universalist that some consider Lewis to be, Lewis is trying to avoid two extremes: on the one hand, the classic universalist position that takes so broad a view that the command to preach the Good News drops out altogether; and, on the other hand, the classic particularist belief that only the relatively few people in the history of our earth who have heard the Gospel and come to believe in Christ are destined to be eternally happy. Lewis believed that the goodness people find in the world religions finds its ultimate source in Christ and that every person will find complete fulfillment in Christ. He also believed that it is possible for a person to reject this fulfillment in whatever form it might come to that person.

[1]See MYTHOLOGY.

UVILAS — One of the great TELMARINE lords in the reign of CASPIAN IX (PC 5). With Belisar, he is "accidentally" killed by arrows at a HUNTING party arranged by the EVIL* MIRAZ.

V

VALLEY OF THE THOUSAND PERFUMES — See MEZREEL.

VANITY — To Lewis's way of thinking, vanity—the desire to be loved beyond the limits God sets—is the chink in a person's armor that allows EVIL to enter in.[1] In the *Chronicles,* pride and vanity are exposed as GREED and love of power. Because LUCY feels ill-favored in the comparison everyone makes between her and SUSAN, she is tempted to know what her friends really think about her and so she uses the spell in the MAGICIAN's book (MN 10). ANDREW is "vain as a peacock" (MN 6), and his wish for power caused him to become a magician in the first place. HWIN discerns that BREE's vanity prevents him from entering NARNIA while his tail is in a shabby condition; Bree, however, calls his vanity "proper respect for [him]self" (HHB 14). It is Andrew's appeal to POLLY's vanity that brings her into his study (MN 1), and Queen Jadis is so vain that she assumes Uncle Andrew has seen a vision of her beauty and has sent the CHILDREN to CHARN to sue for her favor (5).

[1]This is a major theme of Lewis's novel *Perelandra* (1943), in which the unfallen Eve of the planet Perelandra is temped by the Unman through his appeal to her beauty and her position as the mother of her race.

VENGEANCE — See Honor.

VIOLENCE, VIOLENT — Some adult readers find the *Chronicles* too violent and fear that children might, too. As with many men of his generation, especially J. R. R. Tolkien, Lewis's imagination was seared by his experience of trench warfare in World War I. As he used battle imagery in his radio broadcasts during World War II (which later became the book *Mere Christianity*), Lewis uses similar imagery to introduce Aslan in LWW 7, when Mr. Beaver says Aslan has "landed" and is "on the move." The hacking off of the serpent's head makes a gruesome mess on the floor (SC 12), and the description of the battle at Anvard is equally gruesome: horses mauled by cats, one giant shot in the eye, men beheaded (HHB 13). The last battle details the gore: three Calormenes shot by arrows, one pierced by Jewel, one by Tirian's sword, the fox by Eustace's sword; the bull shot in the eye and gored in the side, three dogs killed and one hobbled; and the quiet death of the helpless Bear (LB 11). Violence is also present in other, smaller ways: Puddleglum threatens the Earthman with violence (SC 14); the impact of each giant jumping in rage is likened to the falling of a bomb (6); and Shasta, mistaken for Prince Corin, is promised his first suit of armor and a warhorse for his next birthday (HHB 5).

C. S. Lewis felt that life is violent, and to deny that would be wrong. He was a longtime reader of G. K. Chesterton, who anticipates Lewis's feelings in the essay "The Red Angel":

> a lady has written me an earnest letter saying that fairy tales ought not to be taught to children even if they are true. She says it is cruel to tell children fairy tales, because it frightens them. . . . All this kind of talk is based on that complete forgetting of what a child is like . . . Exactly what the fairy tale does is this: It accustoms [the child] for a series of clear pictures to the idea that these limitless terrors have a limit, that these shapeless enemies have enemies, that these strong enemies of man have enemies in the knights of God, that there is something in the universe more mystical than darkness, and stronger than fear.[1]
>
> [1] *Tremendous Trifles* (New York: Dodd, Mead, 1909).

VOCATION — A call to ADVENTURE from ASLAN to freely obey at deeper and deeper levels of surrender until one's will is transformed into his will. Even Aslan heeds the call he has from the EMPEROR-BEYOND-THE-SEA to obey the Deep MAGIC and the Deeper Magic (LWW 12). PETER is called to fight FENRIS ULF (Maugrim) even if he has no FEELINGS of COURAGE (13). REEPICHEEP—himself under the DRYAD's spell to journey to the Utter East to ASLAN'S COUNTRY (VDT 2)—reminds CASPIAN X that he is not free to abandon his kingship to improperly pursue adventure (16). SHASTA has a profound experience of vocation beginning with his inner complaint about how unfair it is to have suffered so much to get through the desert only to be chased by a lion, continuing with his heart's writhing at being sent without rest by the HERMIT to intercept King LUNE, and climaxing at the revelation by Aslan that he, Aslan, was behind the entire course of Shasta's life up to that point (HHB 11). Aslan calls Eustace and Jill into Narnia and gives them the task of saving Rilian (SC 2). Puddleglum has complete confidence in Aslan's call and providence, even if it may mean his own death (11). Aslan calls Frank and Helen to be king and queen of Narnia (MN 11).

W

WAND — The White Witch's wand is not specifically described; however, it is through her wand that she performs her magic (LWW 2, 4, 11, 13, and 16). Its destruction by Edmund turns the course of battle in favor of Aslan's forces (17).

[Sceptre.]

WAR — See Violence.

WARDEN OF THE MARCHES OF UNDERLAND — The guardian of the borders ("Marches") of Underland and slave of the queen of Underland (SC 9 and 10). As befits a slave of a sunless land, he speaks with a flat voice in a tone as black as night. When he first meets Jill, Eustace, and Puddleglum, he is accompanied by a retinue of one hundred spear-carrying Earthmen. His stock phrase ("Many fall down, but few return") is almost a password, and it allows the warden to deliver the three travelers to the gnomes-in-waiting at the queen's door.

WARDROBE — A freestanding piece of furniture in which to hang and store clothing in the house of Professor Kirke is the means by which the Pevensie CHILDREN first enter NARNIA in LWW. It is fashioned from the MAGIC wood of the apple TREE that grew from the core of the apple of the TREE OF PROTECTION, planted by the young DIGORY. The wardrobe is called "War Drobe" by TUMNUS, who thinks it is a city in SPARE OOM.[1]

[1]Lewis, when a child, had probably read E. Nesbit's story "The Aunt and Amabel" in which Amabel finds her way into a magic world via "Bigwardrobeinspareroom" (250). See LITERARY ALLUSION(S).

WAR OF DELIVERANCE — This term is first used in the GREAT REUNION scene in LB 16 to describe the sometimes VIOLENT civil war in NARNIA that took place 250 years earlier, in 2303 N.Y. The STORY of this war is the chief burden of PC. Serious conflicts break out in the thirteenth year of the rule of the usurper MIRAZ. Prince CASPIAN leads the Old Narnians against the forces of Miraz with the help of PETER and EDMUND. LUCY and SUSAN do not fight; rather, they accompany ASLAN on a series of DANCES: The return of REVELRY frees Narnia from TELMARINE tyranny as much as does the actual combat.

WEREWOLF, WEREWOLVES — In MYTHOLOGY, men who, by will or enchantment, are said to be able to turn into wolves. In LWW 13, werewolves are called to battle by the WHITE WITCH, and some are glimpsed in the first months of the reign of the KINGS and queens (17). In PC 12, NIKABRIK'S unnamed FRIEND has a dull, gray voice that causes PETER's flesh to creep. He is a ravenous, bloodthirsty, hateful beast who urges the HAG to call up the witch and begin the MAGIC. He is killed while in the process of changing from man to wolf as he leaps on CASPIAN and bites him.

WESTERN WILD — An uncivilized country that forms part of the western border of NARNIA (LWW 2 and PC 4). The alpine Western Mountains border a land of high green hills and forests and hide a green valley (LB 16). It is in this valley, with its blue lake, that the GARDEN on the high green hill houses the apple TREE that is so precious to ASLAN (MN 12). No TALKING BEAST lives here; a HUNTER kills a dumb lion here and discards its skin, which is later picked up and used by SHIFT to represent the false Aslan (LB 1).

[GEOGRAPHY, NARNIAN.]

WHITE STAG — A talking male deer, the QUEST of great HUNTING parties of the KINGS and queens of NARNIA, who was said to grant wishes to his captors when captured and, we presume, released (LWW 2 and 17).[1] No physical description of the White Stag is given, but his extraordinary beauty may be inferred from the appearance of the unnamed stag that accompanies DUFFLE in HHB 12. It is portrayed as a strong yet delicate creature with limpid eyes, a dappled coat, and slender legs.

[1]Lewis, as a student of the Middle Ages, would know of the symbolism of the stag for Christ, but it is impossible to determine whether he intended an allusion here.

WHITE WITCH — As the White Witch of LWW, she is responsible for the Hundred Years of Winter that fall on NARNIA; as Jadis in MN, she is the possessor of the DEPLORABLE WORD. When Lewis was writing LWW, he had not yet conceived of MN; when he wrote MN, he strove mightily to show how the White Witch and Jadis are one, but he left loose ends and unanswered questions.

According to Mr. BEAVER, the White Witch is the offspring of a GIANT and the demon LILITH. She is bad through and through,[1] and she is especially villainous because she looks human but has not a drop of human blood in her.[2] In MN Jadis is the last of a long line of apparently human KINGS and queens of CHARN. Lewis's description of the royal statues DIGORY and POLLY encounter at the great hall of Charn is meant quite obviously to show the progression of corruption that occurred in this line, from kindness to cruelty to despair to depravity, culminating in the most EVIL of all, Jadis herself. This corruption is possible only in human beings, as nonhumans (especially the combination of giant and demon) may be assumed to be bad from the start.

At the end of MN, we are told that the TREE OF PROTECTION will keep the witch out of Narnia as long as it flourishes; meanwhile, she has fled to the north, "growing stronger in dark MAGIC." But by the time of LWW, not only has she managed to gain entrance into Narnia, but she has succeeded in gaining enough power over it to hold it in the enchantment of a winter without end. This would imply either that the Tree of Protection has weakened, died, or been gotten around in some other way or that Jadis has become such a master of the dark arts that she is able to overcome the tree's effects. All this must remain speculation, however, because we are never told what happened.

The White[3] Witch — She is the self-proclaimed queen of Narnia, and she is determined to kill the SONS OF ADAM and Daughters of Eve whose thrones, according to the prophecies, are already waiting in CAIR PARAVEL. She rides in a sleigh driven by a DWARF and pulled by white reindeer, and she herself is dressed in white fur to her neck. She wears a gold crown and carries a gold WAND. Her face is deathly pale, and her mouth is crimson. EDMUND thinks her beautiful[4] but stern. She has imposed winter without end on Narnia, and that world is blanketed with a white, killing snow. That she is evil is indisputable: She has no conscience and no scruples and will do anything that is to her advantage. She entraps poor Edmund with the SWEET seduction of TURKISH DELIGHT; she turns Narnians into stone statues that decorate her courtyard; and she sends a pair of wolves to kill the beavers and the CHILDREN. Her worst and most unforgivable offense, of course, is her slaying of ASLAN at the STONE TABLE, during which she endeavors to drive him even deeper into despair by boasting that he has given her Narnia forever, lost his own life, and not saved Edmund after all. Here, however, she shows the VANITY that is in the end her undoing. She assumes that, like her, everyone else lacks a conscience and scruples—and if they do not, then they are fools. In believing she can control everything, she blinds herself to other possibilities. But her knowledge goes only so far: She knows the Deep Magic, but Aslan knows the

Deeper Magic. Her army consists of all the worst ANIMALS and most ghastly creatures of DREAMS and nightmares, including bats, wolves, GHOULS, and HAGS. Aslan's army, powered by the newly freed Narnians, overcomes these beastly apparitions, and the witch herself is killed by Aslan.[5]

Jadis — Jadis is the queen of CHARN, the last of a long line of KINGS and queens, who desires to gain power not only over her own world, but also over as many worlds as possible. Like the White Witch, she is pale and intense. Unlike the White Witch, who is a broadly painted cardboard figure of wickedness, Jadis is more finely drawn. She wears extravagant royal ROBES, bejeweled and embroidered and finer than those worn by any of the other queens in the great hall at Charn. She rules over a dead city, having overcome her sister (the former queen) with the ultimate weapon, the Deplorable Word, which she learned at great cost to herself.[6] Like the White Witch, she also deals in sweet seductions—the maddening verse on the pillar "causes" Digory to strike the GOLDEN BELL, and its sweet tone lets Jadis loose upon the world once more. In ENGLAND, Jadis is truly a vision. In that world she is larger than life, a vision of pure energy on the drab London pavement. Her power, which is sapped by the TIMEless air of the WOOD BETWEEN THE WORLDS, is intensified in our world, and she is able to wrench an iron bar from the LAMP-POST outside the Ketterley residence. That she is a figure of evil is indisputable (witness her selfish destruction of the innocent people of Charn; her merciless beating of Strawberry; her guided tour of the castle, in which she points out to Digory and Polly her favorite spots—the dungeons, the principal torture chambers, and the site of a great and arbitrary slaughter by her grandfather; and her KNOWLEDGE of the Deplorable Word), but she is also an occasion for laughter and the butt of jokes.[7] Her behavior at the lamp-post is outlandish and causes the onlookers to conclude that she is insane; Sarah the maid finds it most exciting. The image of Jadis holding on to Polly's hair as they reenter the Wood between the Worlds is painful, but it also serves to make Jadis seem ridiculous and childish.

Perhaps her most frightening attributes are her desire to come into our world and her assumption that magicians are the ruling class here as well as in Charn. Once she meets Uncle ANDREW, she ceases to notice Digory at all, and Lewis comments parenthetically that "most witches are like that. They are not interested in things or people unless they can use them; they are terribly practical." Finally, she is a symbol to Lewis of the inevitability of the chaos begun by the impersonal experimentation of our technological society.

[COLNEY 'ATCH; FENRIS ULF; PAIN; TECHNOLOGY.]

[1]See DEPRAVITY.

[2]For a discussion of the implications of looking human but being wild, see DUMB BEASTS, n. 1.

[3]The usual meaning of "white witch" as one who practices "white magic" (good magic in which the devil is not invoked) is not the meaning Lewis intends. Here white is used as the color

of death, the time of winter, when nothing can grow and good people are "frozen," prevented from overcoming the evil that holds them in thrall. The stage is thus set for the coming of spring and the rebirth of nature brought by Aslan.

[4]Similarly, Digory thinks Jadis the most beautiful woman he has seen, while Polly doesn't see her beauty at all. This is both an instance of Lewis's unconscious SEXISM—the belief that women are jealous of the beauty of other women—and an ironic comment on the susceptibility of men to the physical beauty of women.

[5]In PC, the HAG informs Doctor CORNELIUS that the White Lady (her name for the witch) is alive and may be called upon for help. Lewis might be suggesting that, though dead herself, the White Witch's evil spirit is able to live on by means of the dark magic.

[6]This is a parallel to Uncle Andrew, who ruined his health by becoming a magician ("One doesn't become a magician for nothing"). It is one of several intentional parallels that Lewis makes in order to impress upon the reader his belief that the quiet violations of natural law in the privacy of a scientist's laboratory are the same as, and lead to, the ultimate violations of the sacredness of life practiced by Jadis and other unscrupulous rulers.

[7]See HUMOR.

WILD FRESNEY — A Narnian wild herb that looks like wood sorrel but tastes nicer when cooked, especially with a little butter and pepper (LB 7).

WIMBLEWEATHER — A small GIANT known as Wimbleweather of Deadman's Hill. He is described as being brave as a lion but not too clever and is a member of CASPIAN's war council. He is a member of EDMUND's parley, a marshall of the LISTS for PETER's single combat, and a major figure in the second battle of Beruna. (PC 7, 13, and 14)

WINE — The spirit of REVELRY is alive in NARNIA, and wine is an important part of celebration. LUCY's healing CORDIAL is probably an alcoholic beverage. In PC 15, a wine list includes thick wines, clear red wines, yellow wines, and green wines. In VDT 6, the crew of the *DAWN TREADER* drinks the strong wine of ARCHENLAND, and in the land of the DUFFERS all but EUSTACE enjoy the mead served (10). The sailors award bottles of wine to the Duffers who win the "boat races" (11), and CASPIAN calls for a round of grog to relax the crew after the escape from the DARK ISLAND (12). On RAMANDU'S ISLAND, REEPICHEEP drinks a toast to the DAUGHTER OF RAMANDU (13). In HHB 5, SHASTA'S CALORMENE meal includes a "white" wine that Lewis says is really yellow. In SC 3, supper at CAIR PARAVEL includes "all manner of wines and fruit drinks." At HARFANG, PUDDLEGLUM (who is fond of the rather strange alcoholic drink of the MARSH-WIGGLES [5]) drinks a saltcellar full of GIANTS' liquor and becomes quite drunk (7). After the DEATH of the QUEEN OF UNDERLAND, RILIAN proposes that they refresh themselves and pledge health to one another with wine left over from their dinner. In MN 6, Uncle ANDREW's fondness for brandy is one of many weaknesses. In LB 2, ROONWIT is offered a bowl of wine by TIRIAN, and the small ANIMALS bring Tirian wine (4). The love of wine, however, reaches the point of excess in SHIFT, who becomes an alcoholic (9).

WOOD BETWEEN THE WORLDS[1] — The wood where POLLY and DIGORY arrive after leaving Uncle ANDREW's study via the MAGIC* RINGS. Polly first calls it the Wood between the Worlds after Digory reasons that they are indeed between worlds. It is so dense and leafy that the light is green, the air is warm, and no sky can be seen. There is no SOUND of life or wind, and the wood is characterized by a FEELING of TIMElessness. There are pools everywhere, and although they look deep they are really shallow. It is hard for the CHILDREN to feel frightened in the wood, and at first it is difficult for them to remember who they are and where they've come from. They feel as if they have always been in the wood. The wood seems to affect people in varying ways, according to their sense of RIGHT AND WRONG. (MN 3, 6, 8, and 15)

[1]A name inspired by William Morris's novel *The Wood beyond the World* (New York: Ballantine Books, 1969).

WOOD PEOPLE — A general class that includes DRYADS, HAMADRYADS, SILVANS, and TREE-PEOPLE. In PC 10, they are called wood GODS and goddesses, and they are the awakened TREES that rush at the TELMARINES. They are certainly the wild people who emerge from the trees when ASLAN commands NARNIA to awake (MN 10).

WOOSES — Haunting spirits, present at the slaying of ASLAN (LWW 14). The term is probably derived from the *Woodwose,* or "wild man of the woods."

WORLD ASH TREE — The expression "the trunk of the World Ash Tree" is found only in the pre-1994 American editions of LWW. According to the WHITE WITCH, the words of the Deep MAGIC are written in at least three sacred places: on the STONE TABLE, on the trunk of the World Ash Tree, and on the SCEPTRE of the EMPEROR-BEYOND-THE-SEA. The written words suggest *runes,* magical letters carved into stone or wood by many early north European cultures. The World Ash Tree is Yggdrasill, the great tree of Scandinavian MYTHOLOGY, a symbol for existence. Its branches tower into the heavens, its trunk upholds the earth, and its three roots reach, respectively, into the realm of the dead, into the land of the GIANTS, and into the abode of the GODS. The holy fountain of fate, the Well of Urd, is under this third root; the council of the gods is convened daily at this fountain. Nearby dwell the three maidens (the Nornir)—Past, Present, and Future (or Fate, Being, and Necessity): They rule the life of every person and the whole world.

[SECRET HILL.]

WORLD'S END — A huge plain of utterly flat land, covered with short green grass. It intersects with the sky, which appears to be a bright blue, glasslike wall. It is located south of the place where REEPICHEEP goes over the last wave of the SILVER SEA. Here EDMUND, LUCY, and EUSTACE meet the LAMB, who is really ASLAN. (VDT 14)

[GEOGRAPHY, NARNIAN; LAST SEA.]

WRAGGLE — The only named SATYR in all of the *Chronicles.* A traitor to the cause of TIRIAN, he fights against him in the LAST BATTLE and loses his life almost immediately when he is struck by one of JILL's arrows. (LB 11)

WRAITHS — Terrifying phantoms, present at the slaying of ASLAN (LWW 14). The word is Scottish in origin and its use is another instance of Lewis's desire to evoke a "northern" atmosphere in his stories.

Y

YOUTH — Despite Lewis's sense of PRIVACY, the adolescents ARAVIS and SHASTA, CASPIAN, EUSTACE and JILL, PETER, SUSAN, and RILIAN live their STORIES in full view of the reader. Thus we see that all but Susan successfully negotiate the crises of late childhood and early adulthood, let us say, from age twelve to age twenty-four.

Lewis is convinced that young people are generous even to the point of being willing to die for something they believe in, that they do most of the serious thinking and worrying they will do before they are fourteen, that they deserve COURTESY from and owe courtesy to the ADULTS in their lives, that "the whole of one's youth [is] immensely important and even of immense length,"[1] and that their appetite for heaven is deeply rooted. On the other hand, youth can be pressured by human respect into conformity and thus greatly tempted by worldliness; in a letter dated 1 August 1953, Lewis tells a friend:

> I . . . think there is lots to be said for being no longer young . . . it is just as well to be past the age when one expects or desires to attract the other sex. It's natural enough in our species, as in others, that the young birds show off their plumage—in the mating season. But the trouble in the modern world is that there is a

tendency to rush all the birds on to that age as soon as possible and then keep them there as late as possible, thus losing all the real value of the *other* parts of life in a senseless, pitiful attempt to prolong what, after all, is neither its wisest, happiest, or most innocent period. I suspect merely commercial motives are behind it all: for it is at the showing-off age that birds of both sexes have least sales-resistance![2]

[AGING AND DISABILITY; CHILDREN.]

[1]Letters 1988, 8 February 1956, 452.

[2]*Letters to an American Lady*, 19.

Z

ZARDEENAH — The moon GODdess to whose service all CALORMENE maidens are dedicated until they marry; before their weddings they offer secret sacrifice to her. Also known as Lady of the Night. (HHB 3)

Appendix One

List of Comparative Ages of Principal Characters in the *Chronicles of Narnia*

All approximate; based on Lewis MS 51, published in *Past Watchful Dragons*, 41–44, and in *The Land of Narnia*, 31.

English People							
Digory	12 in MN,	52 in LWW,					61 in LB
Polly	11 in MN,						60 in LB
Peter		13 in LWW,	27 in HHB,	14 in PC,			22 in LB
Susan		12 in LWW,	26 in HHB,	13 in PC,			21 in LB
Edmund		10 in LWW,	24 in HHB,	11 in PC,	12 in VDT,		19 in LB
Lucy		8 in LWW,[1]	22 in HHB,	9 in PC,	10 in VDT,		17 in LB
Eustace					9 in VDT,	9 in SC,	16 in LB
Jill						9 in SC,	16 in LB
Narnians							
Caspian X				13 in PC,	16 in VDT,	66 in SC	
Rilian						31 in SC	

[1] LWW 5, indicates that Edmund and Lucy are only a year apart in age. Since Lewis MS 51 says that he was born in 1930 and she in 1932, they have to be slightly more than a year apart.

Appendix Two

A Narnian Atlas

(Abridged from the full-size *Companion to Narnia*)

by Stephen Yandell*

We begin our tour with an overview of the Narnian continent (Diagram 1). This expansive body of land is the heart of Lewis's Narnian world, situated in the center of a large encircling ocean that extends to the edges of the world. The small, valley country of Narnia lies on the eastern edge of the continent. As RABADASH reminds the TISROC of CALORMEN, it "is not the fourth size of one of your least provinces" (HHB 8). In the north-east corner of the country

---

*Stephen Yandell, Ph.D., was a Jacob K. Javits Fellow at Indiana University, where he completed his doctoral work in medieval political prophecy. He is an assistant professor of English at Xavier University in Cincinnati, Ohio. The original version of this atlas appeared as "The Trans-cosmic Journeys in *The Chronicles of Narnia*," *Mythlore* 43 (Autumn 1985), 9–23, written when Steve was sixteen years old!

lies LANTERN WASTE (Diagram 2), a forested area that extends from the GREAT WATERFALL and CALDRON POOL to the town of Beaversdam. It is here that Digory and Polly arrive to witness the CREATION OF NARNIA (MN), Lucy appears through the wardrobe (LWW), and Aslan brings Narnia to an end (LB). Equally busy as a point of between-world contact is Narnia's capital of CAIR PARAVEL, located on the eastern border (Diagram 3). It is situated on a peninsula (and for many years an island) where the mouth of the GREAT RIVER meets the great Eastern Sea.

Although a large part of the Narnian continent remains unexplored by Lewis, several prominent regions figure into the *Chronicles*. The WESTERN WILD, for example, is the mountainous region at the center of the continent which contains the GARDEN to which Digory and Polly fly with FLEDGE (MN). West of these mountains lies the country of TELMAR, a nation that invades Narnia in 1998 N.Y. (PC). The northern part of the continent, separated from Narnia by the RIVER SHRIBBLE, consists largely of ETTINSMOOR and the wild wastelands of the North, a mountainous region populated by GIANTS. The south is comprised of ARCHENLAND, a small country separated from Narnia by a single mountain range, and the vast empire of CALORMEN, separated from Archenland by the River Winding Arrow (HHB). The Eastern Sea is dotted with islands, though we learn only of those visited by the *DAWN TREADER*, including GALMA, Terebinthia, the SEVEN ISLES and the Lone Islands (VDT). At the eastern point farthest from the Narnian continent, where the ocean meets the edge of the world, is located the SILVER SEA of lilies and the wave-like opening taken by Reepicheep into

Aslan's country (Diagrams 4, 11). It is here that the Narnian world touches Aslan's country.

Narnia differs from Earth, we're told, because it is a flat world. Caspian X is granted a visit to Earth, in fact, so that he might experience a "round" world for the first time (SC). Lewis's England exists on a globe, while the Narnian continent lies horizontally on a plane which divides the world into distinct upper and lower halves (Diagram 5). Both Earth and Narnia are "round worlds," however, in the sense that each is contained within a large world sphere. The Narnian continent and the ocean are indeed flat, but the sphere surrounding them is round. The occupied areas of the Narnian world are not two-dimensional either. The upper levels of the sphere are inhabited by Stars, living creatures; and the realm of Bism, far below the Queen of Underland's kingdom, makes up the lower part of the sphere. The sphere of the Narnian world thus resembles in many ways a medieval conception of our own world: a series of concentric circles around the world, with a division of all matter between four major elements: earth, air, fire, and water (Diagram 6).

The same form of sphere that contains the Narnian world also encloses the worlds of Earth and Charn, home world of the White Witch. However, the size of each world seems to vary dramatically. The Narnian world is small enough that its edge can be reached over the course of several months of travel, as witnessed in VDT. The vastness of the universe surrounding Earth and its solar system, on the other hand, makes its boundaries undetectable by modern science. What we see of Charn in MN is vast, but too few details are provided to conclude anything about the size of the whole world.

Lewis's cosmology allows for an infinite number of worlds, but only Earth, Narnia, Charn, the Wood between the Worlds, and Aslan's country are named in the Chronicles. The first three are contained within world spheres I have just described, but the final two are infinitely large. Both the Wood between the Worlds and Aslan's country spread out infinitely to accommodate contact with every other world. In MN, Lewis likens the Wood between the Worlds to the cistern corridor, an attic hallway that runs the length of Digory and Polly's London row houses, allowing access to every house. Similarly, the Wood between the Worlds touches the top of each and every world through the infinite pools that dot its landscape (Diagram 7). Individuals travel upward to reach the flat plane of the Woods and travel down through the pools to reach the individual worlds.

Like the hub of a wheel joining multiple spokes, Aslan's country consists of a large central mountain out of which jut an infinite number of mountain ranges (Diagram 8). Within the valley of each range is contained a separate sphere-enclosed world. Earth, Narnia, and Charn, along with countless unnamed worlds, each exists in a valley, separated from one another by large gaps. This geography thus mimics Narnia's own position as a valley country on the Narnian continent. Although the Wood between the Worlds hovers just above the top of the worlds' spheres, it seems to exist in the same space taken up by the higher mountains of Aslan's country. Neither appears to penetrate the other (Diagram 9). The centrality of Aslan's country also reinforces its place as both the geographic and moral center of Lewis's cosmology (Diagram 10).

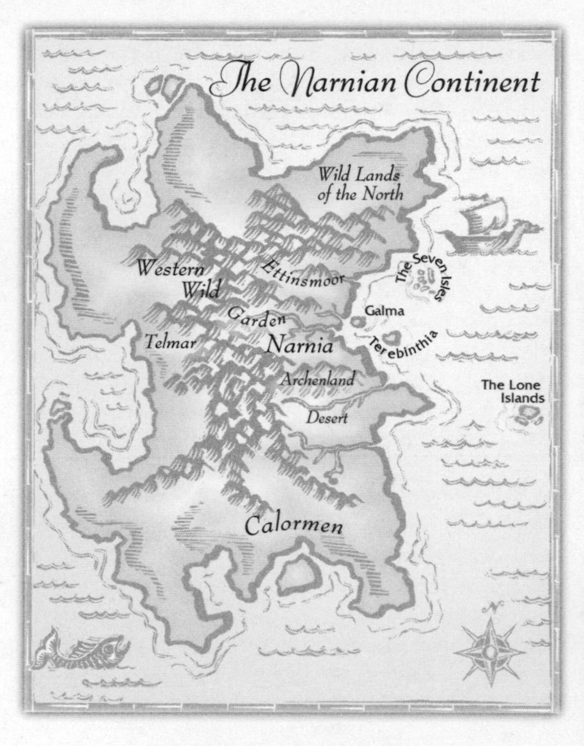

Diagram 1: The Narnian Continent

Diagram 2: Narnia's Lantern Waste

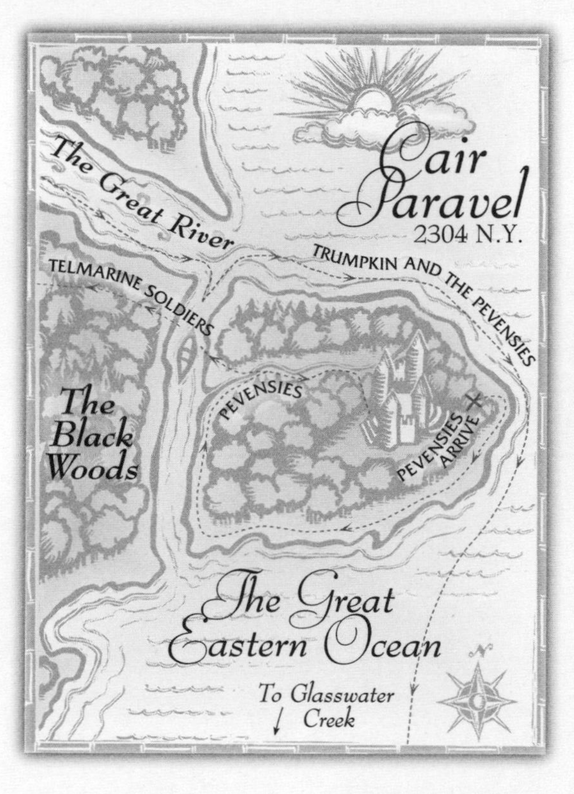

Diagram 3: Narnia's Eastern Peninsula

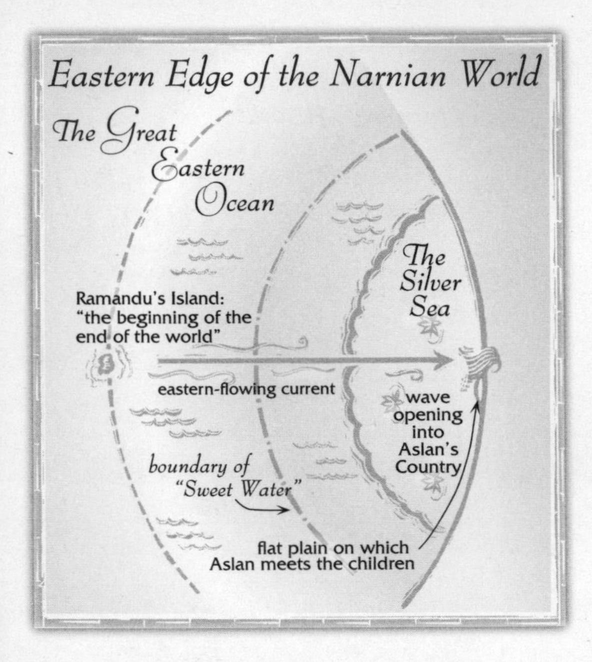

Diagram 4: Eastern Edge of the Narnian World

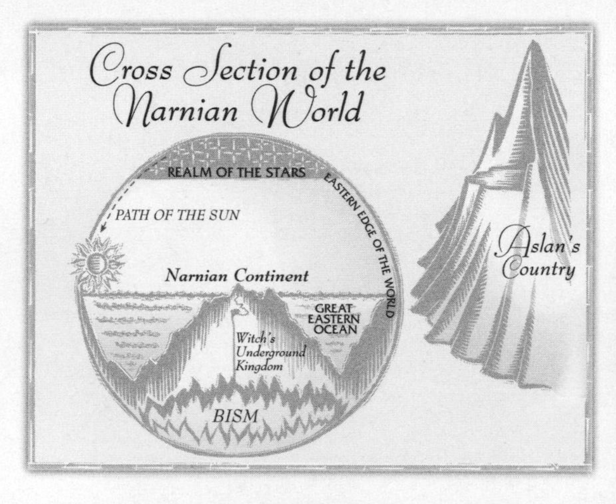

Diagram 5: Cross Section of the Narnian World

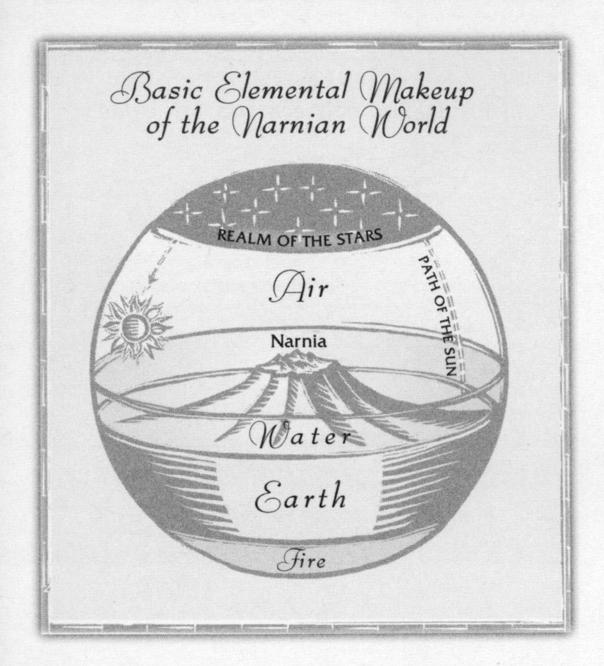

Diagram 6: Basic Elemental Makeup of the Narnian World

Diagram 7: The Wood between the Worlds

Diagram 8: Aslan's Country and the Surrounding Worlds

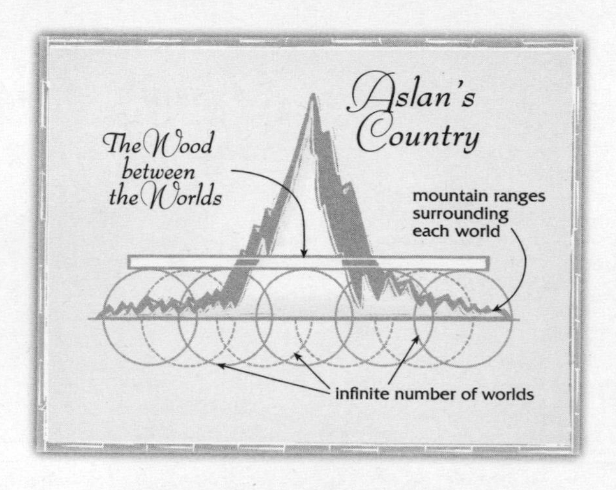

Diagram 9: Aslan's Country in Relation to the Wood between the Worlds

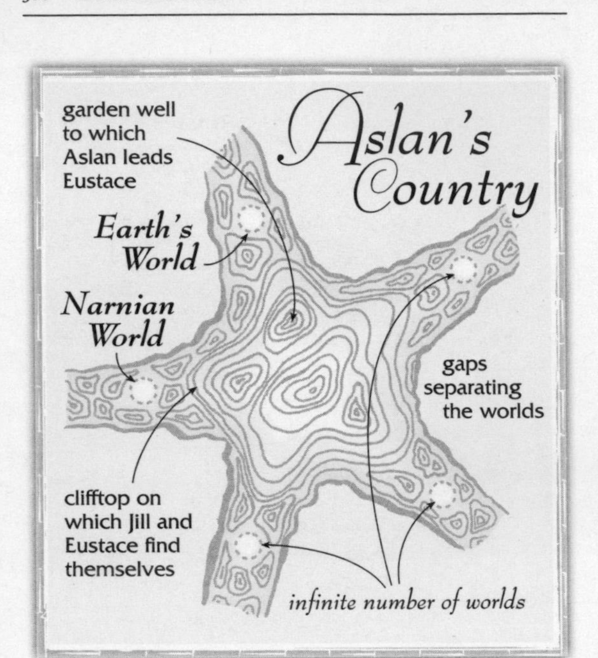

Diagram 10: Aslan's Country, Topography

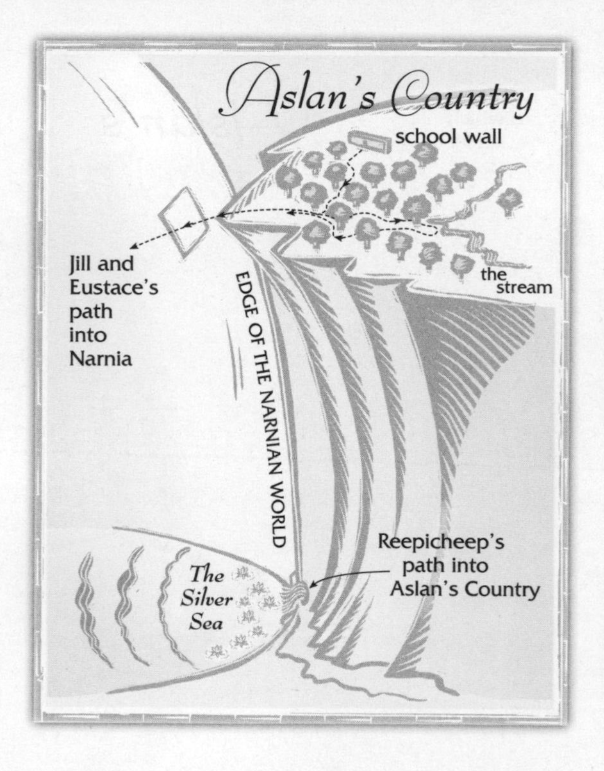

Diagram 11: Aslan's Country Meets the Narnian World

Where You Will Find More Information in the Full-Size Companion to Narnia

When you have exhausted the contents of this *Pocket Companion,* you might want to graduate to the full-size *Companion to Narnia.* Why?

It's almost twice as long, has more diagrams and charts, and has much more information about all of the major characters and themes.

The great author Madeleine L'Engle wrote a wonderful Foreword.

I say a lot more in the Introduction about how to read the *Chronicles of Narnia,* about what Lewis was writing and why, about what is a story, and about the order in which the *Chronicles* should be read. I end with a personal note about what the books mean to me.

I think you'll want to get the full-size *Companion* to read a book-by-book discussion of Aslan. In the *Pocket Companion* I just write 538 words about Aslan, but in the full-size

Companion I write 10,679 words. I show his connection to every major character and theme.

I can't guarantee you'll find twenty times more information about Aravis, Bree, Caspian, Digory, Edmund, Eustace, Hwin, Jill, Polly, Puddleglum, Reepicheep, Rilian, Shasta, Susan, and the White Witch, and nearly three hundred more people, places, things, and themes that are in the full-size *Companion*. But you will learn a lot more.

Thank you for letting me be your companion to Narnia.

Acknowledgments

"The test of all happiness is gratitude," said G. K. Chesterton. And I am a very happy man because I have the opportunity to thank all the people who have helped me write the *Pocket Companion to Narnia.*

I am in deepest debt to Joshua Falconer and Rebecca King Cerling. They read the fourth and fifth editions of *Companion to Narnia* attentively and made many helpful suggestions for improvements. Joshua changed every page reference to every English edition of the *Chronicles* into chapter references for the *Pocket Companion to Narnia*. He also made all the editorial changes, which left me free to attend to the content of the books. Becky (with the encouragement of her husband and my friend, Lee, and their children, Ella and Brendan) made all the recommendations for reworking the *Companion* to be more "kid-friendly" in the *Pocket Companion.* John Marheineke, a new (and, I hope, lifelong) companion to Narnia, made many more improvements. Elizabeth Patterson and Lee Cerling proofread the typescript and offered many suggestions.

For the *Pocket Companion to Narnia* I assembled a company of astute fifth-, sixth-, seventh-, and ninth-grade readers: my nephews and niece—David, Emily, and Paul (my namesake)—and their parents, Raymond and Laura, Nathan Blackmon (and his father, my dear friend Rick), Erik and John Lake (and brother Thomas, and their mother and father, Silvia and Don, my splendid friends), Sarah Shadduck (and Catherine and Bob, her mom and dad), and Robert Williams (and his mother, Maura). Without them I would not have been realistic in the *Pocket Companion*.

I am especially grateful to my first Harper & Row editor, Roy M. Carlisle, who came up with the idea for the *Companion*. Roy encouraged me to find an artist who would try to do in our generation what Pauline Baynes did in hers: I found this artist in Lorinda Bryan Cauley (and Baynes agrees).

I thank my new editor, Gideon Weil, at HarperSanFrancisco. I am also grateful to production editor Lisa Zuniga, copyeditor Karen Stough, and designer Ralph Fowler.

About the Author

Paul F. Ford has been a student of the life and writings of C. S. Lewis since 1961. He received his master's degree with his thesis "The Life of the World to Come in the Writings of C. S. Lewis" and his doctorate with his dissertation "C. S. Lewis: Ecumenical Spiritual Director: A Study of His Experience and Theology of Prayer and Discernment in the Process of Becoming a Self." He founded the Southern California C. S. Lewis Society in 1974 and was vice-president of the C. S. Lewis Foundation, Redlands, California. He is also a director of the Kilns Oxford, Ltd., and was instrumental in the purchase and renovation of Lewis's home as a study and retreat center.

Dr. Ford studied for the Roman Catholic priesthood for twelve years. Rather than being ordained, he became a Benedictine monk at St. Andrew's Abbey, Valyermo, California, where he lived for five years. The monastery sent him to be the first Roman Catholic doctoral student in the School of Theology at Fuller Seminary, Pasadena, California, where, for ten years, he was a teaching assistant and lecturer in New Testament spirituality and in the theology of C. S. Lewis. He was the scholar guest of honor of the

Mythopoeic Society for its observations of the centenary of Lewis's birth at Wheaton College, Wheaton, Illinois, in the summer of 1998.

Dr. Ford contributed eleven entries, including all the entries on the *Chronicles of Narnia,* to *The C. S. Lewis Reader's Encyclopedia,* Jeffrey D. Schultz and John G. West Jr., eds. (Grand Rapids: Zondervan, 1998).

Dr. Ford is a professor of theology and liturgy at St. John's Seminary in Camarillo, California. He is married to the philosopher Janice Daurio, Ph.D., who teaches philosophy at Moorpark College. They make their home in Camarillo.

As a leader in worship education, Dr. Ford was honored by Cardinal Roger Mahony with the Laudatus Award for 1995 "for excellence in the promotion of the liturgical life of the parishes and the people of the Archdiocese of Los Angeles."

Dr. Ford's latest music publications are the book and CD *By Flowing Waters: Chant for the Liturgy* (1999), the booklet and double-CD *Lord, By Your Cross and Resurrection: The Chants of By Flowing Waters for Holy Week and Easter Sunday* (2001), and *Psallite: Sacred Songs for Liturgy and Life* (CD, music book, full music edition, 2005)—all published by The Liturgical Press, Collegeville, MN.

Updates to this *Pocket Companion* and to the full-size *Companion* and other Narnian helps are available at his website: www.pford.stjohnsem.edu/ford/cslewis/narnia.htm. Contact him at: companion_to_narnia@hotmail.com.